Praise for *Critique of Intelligent Design*

Finally we have a book on so-called 'intelligent design' that gets to the heart of the matter rather than devoting all its energies to a point by point refutation of that doctrine. While providing a sophisticated modern understanding of the complexities of organisms and the biological processes that have resulted in life as it has evolved, the authors of *Critique of Intelligent Design* never lose sight of the real issue which is the struggle between materialism and supernaturalism as an explanation for the world of phenomena. Theirs is the model on which all discussions of intelligent design should be based.

> —RICHARD LEWONTIN, Alexander Agassiz Research Professor at the Museum of Comparative Zoology, Harvard University; co-author of *Biology Under the Influence* (Monthly Review Press) and *The Dialectical Biologist* (both with Richard Levins)

The intelligent design creationist movement's attack on the natural sciences has been thoroughly critiqued and rightly rejected. However, the movement's attempt to undermine the social sciences as well has been largely overlooked. This book fills that void by offering a thoughtful, well-researched discussion of the major figures—besides Darwin himself—whom ID creationists demonize: Epicurus, Marx, and Freud. Moreover, by analyzing C. S. Lewis's influence on the ID movement's leaders, the authors further expose ID as essentially an exercise in Christian apologetics. This is an excellent book. It adds to the growing body of critical writing about intelligent design creationism.

> —BARBARA FORREST, a key witness on the side of evolution in the landmark Dover, PA trial; professor of philosophy, Southeastern Louisiana University; co-author of *Creationism's Trojan Horse: The Wedge of Intelligent Design* (with Paul R. Gross)

A discerning historical reconstruction, which succeeds in illuminating the broad anti-materialist agenda underlying the intelligent design movement.

> —DAVID SEDLEY, Laurence Professor of Ancient Ph?l?   ?rsity of Cambridge, UK; author of *Cr?*

A scholarly and compelling bo?   anti-Enlightenment project—and on   ?dden reactionary agendas. Anyone in   than

fairy tales about a Celestial Designer should get hold of a copy. So too should educators intending to force intelligent design onto their pupils.

—PETER DICKENS, Faculty of Social and Political Sciences, University of Cambridge, UK; author of *Society and Nature*

Debates about religion and science are back on the table. After numerous attacks on modern science from the creationist or 'intelligent design' side, there has been a counterattack coming from a few natural scientists and materialist philosophers. What was missing and what this book provides is an enlightened Marxist perspective. Through a clear presentation of the works of Epicurus, Lucretius, Hume, Feuerbach, Marx, Darwin, Freud, Lewontin, and Gould, as well as of their adversaries, the authors provide a fascinating history of the long struggle between scientific and materialist thought on the one hand and religion and various forms of idealism on the other—probably the most significant issues over which humans have been arguing throughout their recorded history.

—JEAN BRICMONT, professor of theoretical physics, University of Louvain, Belgium; author of *Humanitarian Imperialism* (Monthly Review Press) and co-author of *Fashionable Nonsense* (with Alan Sokal)

With Epicurus and Darwin among its heroes, this book is a timely exposure of the creationist dogmatism that the intelligent design movement seeks to disguise as science.

—A. A. LONG, professor of Classics and Irving Stone Professor of Literature, University of California, Berkeley; author of *Hellenistic Philosophy: Stoics, Epicureans, Sceptics*

In combating the new creationism repackaged as intelligent design, it is not enough to refute particular misunderstandings about chance, complexity, or natural selection. ID is part of an offensive against materialism and humanism aimed at imposing a Christian fundamentalist culture congruent with the needs of a declining empire. *Critique of Intelligent Design* places the debate in its broadest context and historical roots from Epicurus on up, in a vigorous defense of a materialist view of nature that rejects the tepid compromise that would simply divide the turf into domains of science and religion.

—RICHARD LEVINS, John Rock Professor of Population Science, Department of Population and International Health, Harvard University; co-author of *Biology Under the Influence* (Monthly Review Press) and *The Dialectical Biologist* (both with Richard Lewontin)

CRITIQUE *of* INTELLIGENT DESIGN

Materialism versus Creationism from Antiquity to the Present

John Bellamy Foster, Brett Clark, and Richard York

MONTHLY REVIEW PRESS
New York

Library of Congress Cataloging-in-Publication Data
Foster, John Bellamy.
Critique of intelligent design : materialism versus creationism from
antiquity to the present / John Bellamy Foster, Brett Clark, and Richard
York.
    p. cm.
 ISBN 978-1-58367-173-3 (pbk.) -- ISBN 978-1-58367-174-0 (hardback)
 1. Intelligent design (Teleology) 2. Creationism. 3. Materialism. I.
Clark, Brett. II. York, Richard. III. Title.
 BS651.F715 2008
 146'.3--dc22

                        2008036623

Monthly Review Press
146 West 29th Street, Suite 6W
New York, NY 10001

5 4 3 2

CONTENTS

Preface

The intelligent design movement, which arose in the United States in the 1990s and quickly obtained headlines through its challenge to the teaching of evolution in the public schools, sees itself as the outgrowth of a 2,500-year *critique of materialism* dating back to the ancient Greeks. Our intent in this short book is to look at this same debate from the opposite point of view, by providing a brief account of the 2,500-year materialist *critique of intelligent design* (creationism) out of which the modern scientific worldview emerged. This millennia-long controversy within Western thought is examined as it bears on social science as well as natural science, philosophy as well as religion, and the state (politics) as well as the church.

Numerous recent attempts to respond to the intelligent design movement have sought to forge an artificial peace between science and religion. Yet the conflict between religion and science, which the intelligent design movement brought to the fore, is, we will contend, insurmountable within the present society. Religious alienation, i.e., *alienation from the world*, is a reflection of human alienation, as is the alienation of science, when conceived as a mere instrument of domination. Both are equally necessary to the present structure of power. The only way to transcend this dual estrangement is to create through social means a broader material-ism-humanism, in which a sustainable relation to nature, i.e., a lived naturalism, is the first precondition. To achieve this, however, we will have to change our relation to the world, making it *our friend*.

This book, more even than most, is a social product, involving our family, friends, and colleagues. Our argument, which developed over several years, had its first manifestation in a public lecture by Brett Clark, delivered at a number of universities in 2005–6, and then in an article, co-authored by the three of us, in *Theory & Society* in 2007.[1] We would like to thank the editors of *Theory & Society*, especially Karen Lucas, for their help and support at this earlier stage of work. We are also grateful to our friends at Monthly Review for their encouragement, including John Mage, Martin Paddio, John Simon, Michael Yates, Claude Misukiewicz, and Scott Borchert. It is impossible to imagine the present work apart from the inspiration offered by Carrie Ann Naumoff, Kris Shields, and Theresa Koford, through their own struggles on behalf of education, human welfare, and life in general—from which this book derives much of its practical meaning. Finally, we would like to acknowledge, in the persons of Saul and Ida Foster and Arthur and Galen York, a generation of students in public education, to all of whom, and to the hope for the world that they represent, this book is dedicated.

1. Introduction

It is one of those ironies that dot the course of history that the oldest known use of the term "intelligent design" in today's sense can be traced to a letter written by Charles Darwin to John Herschel in May 1861, questioning the notion that the world was designed. Darwin wrote, "The point which you raise on intelligent Design has perplexed me beyond measure. . . . One cannot look at this Universe with all living productions & man without believing that all has been intelligently designed; yet when I look to each individual organism, I can see no evidence of this. For I am not prepared to admit that God designed the feathers in the tail of the rock-pigeon to vary in a highly peculiar manner in order that man might select such variations & make a Fan-tail."[1]

For Darwin the notion of "intelligent design" was the subject of critique. Yet, since the 1990s this same term has been resurrected by group of creationist thinkers in the United States in an attempt to challenge materialism and Darwinism, and to wedge fundamentalist Christian views into science and culture. This new "intelligent design" movement has sought to provide "scientific" evidence of design in nature, with the object of transforming the entire basis of public life.

The initial point of attack, because in many ways it is the weakest link, has been the public school system. Today's proponents of "intelligent design," the "creation science" of earlier decades in a new guise, are reigniting the age-old war between materialism and creationism by attempting to elevate their doctrine to the level of science and to incorporate it as part of the science curriculum in public schools—to be given equal standing with evolutionary theory.[2]

High school textbooks at the end of the nineteenth century had incorporated many of the insights of evolutionary theory. By the beginning of the twentieth century, the major textbooks related to biology, botany, zoology, and geology revealed "a strong evolutionary flavor" and excluded or failed to endorse "creationist concepts."[3] Teachers were encouraged and instructed to teach evolution. At the same time, a massive expansion of secondary education increased the number of students being exposed to evolutionary theory. By the 1920s, Christian conservatives sensed a threat and mobilized an anti-evolutionary crusade. At the forefront of this crusade, William Jennings Bryan, Woodrow Wilson's secretary of state and three-time Democratic Party presidential candidate, proclaimed the necessity to "drive Darwinism from our schools."[4] A famous orator, known particularly for his "Cross of Gold" speech that affected millions, Bryan had by the 1920s become in the words of H. L. Mencken "a tinpot pope in the Coca-Cola belt," promoting Christian fundamentalism. In 1923, after years of fundamentalist organizing, Oklahoma passed the first anti-evolution law. Later that year, Florida passed a similar resolution. Then in 1925 Tennessee followed suit, outlawing the teaching of Darwinism and any theory that denied divine creation as presented in the Bible. Known as the Butler Act it declared it "unlawful for any teacher in the Universities, Normals and all other public schools in the state . . . to teach any theory that denies the story taught in the Bible, and to teach instead that man has descended from a lower order of animals." This set the stage for the epoch-making Scopes Trial in Dayton, Tennessee—better known as the "Scopes Monkey Trial."

Eager to challenge the Tennessee statute, the American Civil Liberties Union offered free counsel. Looking to make a name for their town, Dayton business leaders gathered together at a drug-store meeting and decided to challenge the law. John Scopes, a high school physics instructor and athletics coach, who, while teaching biology as a substitute for the regular instructor, had assigned readings on evolution from the class textbook (*A Civic Biology* by George William Hunter) as part of a review for an exam, volunteered to be arrested for breaking the statute. The court case that ensued pitted Clarence Darrow (attorney for the defense) against William Jennings Bryan (attorney for the prosecution). For eight days in 1925, journalists, visitors, and Dayton townspeople were swept up in the fury, watching the duel of the giants. Bryan railed against evolution both inside and outside the courtroom, proclaiming that many social ills were tied to the teaching of evolution. Darrow called Bryan to the stand, questioning the latter in regard to the literal truth of the Bible as it related to the world as it was known. Darrow showed that the stories of the Bible—such as the sun standing still, which Bryan accepted as truth—were unreasonable and wrong in light of scientific knowledge and that they should not be used as the basis for teaching science. While Darrow devastated Bryan on the stand, Scopes was found guilty of teaching evolution (the decision was later overturned on a technicality). Bryan died in Dayton five days after the trial ended.[5]

Despite the attention generated by the Scopes Trial, evolution almost completely disappeared from public classrooms in the United States in the decades that followed. Although never enforced, the Butler Act remained on the books and was not repealed until 1967. The Soviet launching of the *Sputnik* satellite in 1957 and the space and nuclear arms races eventually resulted in the perception of an educational crisis in the United States and led to a refurbishing of science education in particular. This included the renewed teaching of evolutionary theory. Updated biology books for high schools incorporated sections on evolution. Immediately, the anti-evolutionary movement mobilized, proposing "scientific creationism." Under this guise, a few states, including Arkansas and Louisiana, adopted acts that required "equal time" teaching evolution and creationism within classrooms.

In 1982, in *McLean v. Arkansas Board of Education*, a federal court ruling declared that statues aimed at "balanced treatment" between evolution-science and so-called creation-science violated the Establishment Clause of the U.S. Constitution—"Congress shall make no law respecting an establishment of religion, or prohibiting the free exercise thereof"; that "creation-science" was not in fact science. In 1987, in *Edwards v. Aguillard*, a group of scientists, including Nobel laureates, submitted a brief in support of the individuals who were challenging the constitutionality of Louisiana's "equal time" act. The brief pointed out that science is devoted to investigating natural phenomena and providing naturalistic explanations. In other words, a commitment to materialism is at the foundation of science. The U.S. Supreme Court ruled in their favor, protecting the teaching of evolution in public schools, and declared that by advancing the notion that a supernatural being had created humankind, "creation science" impermissibly endorsed religion. Other court defeats for creationism occurred in 1990, 1994, and 1997.[6]

Undeterred, anti-evolutionists simply regrouped for their next crusade, hoping to find a crack in the legal and political apparatus in which to wedge theology and to undermine evolutionary theory. As the millennium approached, Carl Sagan reflected that reactionary forces were at work and constant vigilance was required to defend science. He noted:

> I worry that, especially as the Millennium edges nearer, pseudoscience and superstition will seem year by year more tempting, the siren song of unreason more sonorous and attractive. Where have we heard it before? Whenever our ethnic or national prejudices are aroused, in times of scarcity, during challenges to national self-esteem or nerve, when we agonize about our diminished cosmic place and purpose, or when fanaticism is bubbling up around us—then, habits of thought familiar from ages past

reach for the controls. The candle flame gutters. Its little pool of light trembles. Darkness gathers. The demons begin to stir.[7]

And stir they did, as the forces of anti-evolution repackaged their dogma, promoting "intelligent design" (a reworked version of pre-Darwinian natural theology) as science in an attempt to appeal to common prejudices in an alienated world.

According to the intelligent design school text, *Of Pandas and People*,

> Darwinists object to the view of intelligent design because it does not give a natural cause explanation of how the various forms of life started in the first place. Intelligent design means that various forms of life began abruptly, through an intelligent agency, with their distinctive features already intact—fish with fins and scales, birds with feathers, beaks, and wings, etc.[8]

Taking a page from the civil rights movement, intelligent design proponents insisted that a diversity of ideas should be taught in schools. "Teach the controversy," they proclaimed, attempting to cast doubt on the legitimacy of evolutionary theory, and suggesting that it sought to cut off legitimate scientific inquiry based on a dogmatic adherence to metaphysical materialism/naturalism. Claiming that they wanted to expand scientific inquiry through consideration of intelligent design, they advocated a position that had as its ultimate goal the overthrow of science. The foremost advocate for the intelligent design view, complaining of discrimination against design notions, was not a biologist but a legal scholar, Phillip E. Johnson, who almost single-handedly launched the intelligent design movement with his 1991 book *Darwin on Trial*.[9]

In recent years anti-evolutionists have sought to force the teaching of intelligent design through elected school board officials and the courts. Although the teaching of evolution is mandated, this does not translate into the actual teaching of evolution, as many teachers ignore or play down the importance of evolution to avoid confrontation and controversy. Many states, such as Kansas and Ohio, have been the sites of concerted efforts by intelligent design proponents to establish new teaching standards that further undermine the teaching of evolution. Such attempts have been contested and resulted in the rallying of the public on both sides.

The most noteworthy event commenced in Dover, Pennsylvania, in 2004, where the Thomas More Law Center—a religious organization eager to promote intelligent design and to challenge the American Civil Liberties Union in court—found members on the Dover school board willing to promote the teaching of intelligent design. In October 2004, the school board voted to change the biology curriculum, noting that students should be made aware of the gaps and problems in Darwin's theory. Intelligent design supporters argued that evolution was "just a theory" and that facts did not support the theory. Intelligent design was actively promoted as an alternative to evolution. Teachers were ordered to read a statement questioning evolution and making a case for intelligent design to all ninth grade biology students. Copies of the intelligent design textbook, *Of Pandas and People*, which promoted this statement, were given to the school. The teachers as a group refused to read the statement to their students. In December 2004, parents in Dover filed suit against the district in the case *Kitzmiller et al. v. Dover Area School District* (the case is commonly referred to as the Kitzmiller or Dover trial). The trial ran from September 26 to November 4, 2005. A George W. Bush appointed judge, John E. Jones III, a conservative Christian, presided.[10]

The supporters of intelligent design put together a group of expert witnesses they hoped would give credibility to their position. However, they were hindered in this by the fact that virtually no serious scientists were willing to support intelligent design. A reluctant Discovery Institute—the Seattle-based center for the promotion of intelligent design, with which nearly all of its leading proponents are associated—participated in the trial. Michael Behe, a senior fellow at the Discovery Institute, attempted in his testimony to separate intelligent design from religion, claiming that intelligent design was founded on empirical evidence showing irreducible complexity that could only be attributed to design. Behe had defined irreducible complexity in his 1996 book *Darwin's Black Box* and again in his 2001 article "Reply to My Critics" as

> a single system which is composed of several well-matched, interacting parts that contribute to the basic function, wherein the removal of any one of the parts causes the system to effectively cease functioning. An irreducibly complex system cannot be produced directly (that is, by continuously improving the initial function, which continues to work by the same mechanism) by slight, successive modifications of a precursor system, because any precursor to an irreducibly complex system that is missing a part is by definition non-functional.... Since natural selection can only choose systems that are already working, then if a biological system cannot be produced gradually it would have to arise as an integrated unit, in one fell swoop, for natural selection to have anything to work on.[11]

In cross examination, Behe had to admit that there was not a single peer-reviewed publication within science supporting his view of irreducibly complex biological systems that could not possibly be accounted for by evolution. Behe himself had admitted in his "Reply to My Critics" that pointing to an irreducibly complex system (where a missing part would cause the system to cease to function) was not the same as proving that such irreducible complexity could not arise through natural selection. In evolutionary theory the concept of exaptation is widely used to account for the evolution of a biological feature that serves a function at one point in evolution and switches to another function later on due to changes in the subject system. With this concept biologists have been able to explain the gradual evolution of complex systems such as the human eye, which can thus be shown not to have arisen in "one fell swoop." To make matters worse, Behe had described (1) the bacterial flagellum, (2) the blood-clotting cascade, and (3) the immune system as three irreducibly complex systems. Yet, Kenneth Miller, a Brown University biology professor and specialist in cell biology, presented solid evidence at the trial that each of these systems had been shown in peer-reviewed scientific research not to be irreducibly complex. Moreover, even if such negative arguments with respect to current evolutionary theory were proven to be correct, as Judge Jones himself noted, this would not have in any way constituted positive evidence of design, which was beyond scientific proof.[12]

Steve Fuller, a sociologist and philosopher of science, also testified in support of intelligent design, arguing that presenting design arguments in science classrooms would improve science education. He went so far as to suggest that there ought to be a kind of affirmative action program in support of intelligent design. Fuller openly acknowledged that intelligent design was a form of creationism—something that others on the intelligent design side in the trial had sought to deny.

On the evolutionary side, philosopher Barbara Forrest, co-author of *Creationism's Trojan Horse: The Wedge of Intelligent Design*, provided damning evidence in the trial of how intelligent design was simply creationism in disguise. She revealed how throughout *Of Pandas and People* "creation" was used in early drafts (1983 through 1987). Later the word *creation* was expunged from the working draft, only to be replaced with the phrase "intelligent design." The new phrase was employed in an effort to circumvent recent court rulings on religion, science, and public schools. Even the definition of "intelligent design" in *Of Pandas and People* was the same as the definition of "creation" in the earlier drafts of the book. The original title of the book itself was *Creation Biology*.

John Haught, a Catholic theologian from Georgetown University, testified on the evolutionary side that intelligent design arguments were not new but dated back to Thomas Aquinas and

even earlier, and had been advocated most famously by natural theologian William Paley at the beginning of the nineteenth century.
This was agreed to by the intelligent design proponents, although
unlike Paley they did not "officially" designate the Designer as God,
in a dubious attempt to separate their "science" and theology.

In the end Judge Jones, in a 139-page decision, determined that
intelligent design was not science. It was "nothing less than the
progeny of creationism. . . . [T]here is hardly better evidence of
ID's [intelligent design's] relationship with creationism than an
explicit statement by defense expert Fuller that ID is a form of creationism. . . . The goal of the IDM [intelligent design movement] is
not to encourage critical thought, but to foment a revolution which
would supplant evolutionary theory with ID." Viewing the intelligent design claims to "science" as constituting little more than traditional creationist "God of the gaps" arguments (God exists wherever gaps in scientific evidence exist), Judge Jones declared that its
promotion in public schools violated the Establishment Clause of
the U.S. Constitution.

Despite the intelligent design movement's resounding defeat in
the Dover trial, the holy war continues. Anti-evolutionary plat-
forms have emerged in numerous states (Alaska, Iowa, Minnesota,
Missouri, Oklahoma, Oregon, and Texas). Intelligent design proponents continue to mobilize, creating their own journals, maga-
zines, and Web sites, attempting to bolster their scientific appearance. They have hired public relations firms and have tried to work
the media to establish a position within the fourth estate. The intelligent design cause is bolstered by the widespread belief in creationist views among the U.S. population, and by the growth of
Christian fundamentalism and conservative Christian views as a
whole. A 2005 cross-national survey found the United States was
thirty-third among thirty-four industrial nations (just above
Turkey) in the percentage of the population that accepted human
evolution. Incredibly, a full 45 percent of the U.S. population,
according to a series of Gallup polls, believes human beings were

created in their present form sometime in the last 10,000 years, while much of the remainder of the population is only tentative in its support of human evolution. Exploiting the prevalence of anti-evolution views, the leadership of the Discovery Institute seeks nothing less than to transform the entire culture by overturning not only evolutionary theory but scientific materialism as a whole. Indeed, despite official denials, the intelligent design movement is almost completely devoted, as its proponents make clear in innumerable statements, to the promotion of a conservative or fundamentalist Christian worldview.[13]

The intelligent design movement's goals can be described as more theological than scientific, more political than theological. As C. S. Lewis, the patron saint of the intelligent design movement, wrote in the preface to *The Great Divorce*: "But what, you ask, of earth? Earth, I think, will not be found by anyone in the end to be a very distinct place. I think earth, if chosen instead of Heaven, will turn out to have been, all along, only a region in Hell; and earth, if put second to Heaven, to have been from the beginning a part of Heaven itself."[14] In this conception, the earth becomes indistinct, virtually disappears, along with nature and humanity, and the only realities that remain are Heaven and Hell, i.e., the Day of Judgment. There is no doubt that Christian apologetics of this kind are central to the intelligent design movement.

Even more important than the religious aspect of the new creationism, however, is its political aspect. Writing in the early 1980s Stephen Jay Gould went so far as to contend that "the core of practical support" for today's creationism "lies with the evangelical right, and creationism is a mere stalking horse or subsidiary issue in a political program that would ban abortion, erase the political and social gains of women by reducing the vital concept of the family to an outmoded paternalism, and reinstate all the jingoism and distrust of learning that prepares a nation for demagoguery."[15]

Nevertheless, the theological and political aspirations of today's intelligent design movement should not erase the fact that crucial issues about the historical-dialectical development of science and religion, materialism and creationism, evolution and design are raised by this controversy. Western science itself is a product in large part of a 2,500-year critique of intelligent design that was tied to larger social struggles occurring over the same vast period.

Materialism versus Religion

In order to understand the full historical significance of the intelligent design movement it is crucial to recognize that it sees as its enemy not simply Darwinian evolutionary theory, but materialist science (including social science), philosophy, and culture in general. In this way it is part of a millennia-long debate over materialism and design stretching back in Western society to the ancient Greeks. (In this book we will not look beyond the strictly Western debate over science and religion.) Indeed, the critique of materialism by design thinkers has had its counterpart in the critique of intelligent design by their materialist opponents. This debate, which is older than Christianity, had, from the first, political and social, as well as scientific and religious, ramifications.

On a philosophical level, the key defining feature of the emerging scientific worldview has always been a commitment in some sense to materialism (sometimes also called naturalism), i.e., the view that the world is explained in terms of itself, by reference to material conditions, natural laws, and contingent, emergent phenomena, and not by the invocation of the supernatural. This commitment stems from the intellectual foundation laid by the Greek atomists Democritus and Epicurus over two millennia ago, which helped inspire the European scientific revolution in the sixteenth and seventeenth centuries. It is the fundamental incompatibility of thoroughgoing materialism with a teleological or religious worldview—insofar as they each attempt to account for *natural* phenomena—that has driven the conflict between science and religion from antiquity to the present.

In challenging the teleological worldviews of Socrates, Plato, and Aristotle, Epicurus at the beginning of the third century BCE drew on the tradition of Greek atomism (Leucippus and Democritus) to banish the gods from the world and to build a framework for understanding nature and society free from superstition. As Epicurus's Roman follower, the poet Lucretius, expressed it (in the first century BCE) in *De rerum natura* (literally *On the Nature of Things*):

> Therefore this terror and darkness of the mind
> Not by the sun's rays, nor the bright shafts of day,
> Must be dispersed, as is most necessary,
> But by the face of nature and her laws.

> We start from her first great principle
> That nothing ever by divine power comes from nothing.
> For sure fear holds so much the minds of men
> Because they see many things happen in earth and sky
> Of which they can by no means see the causes,
> And think them to be done by power divine.
> So when we have seen that nothing can be created
> From nothing, we shall at once discern more clearly
> The object of our search, both from the source from which each thing
> Can be created, and the manner in which
> Things come into being without the aid of gods.[16]

Epicurean materialist philosophy faded from the intellectual landscape with the spread of Christianity and the ensuing decline of reason and secular learning in the medieval era. However, a revival in materialism began in the fifteenth century, as Lucretius's great poem was rediscovered and began to circulate widely, inspiring thinkers engaged in the scientific revolution of the early modern era and leading to Gassendi's systematization of Epicurean natural philosophy in the seventeenth century. As historian of science David Lindberg has recently argued, it was this revival of materialism, rather than the emergence of experimental methods and mathematical advances, that led to the scientific revolution of the sixteenth and seventeenth centuries and ultimately to the Enlightenment.[17]

Throughout this book, we will focus on materialism *as it emerged over the course of history through the critique of intelligent design*. Materialism in this sense then becomes the defining feature of science and indeed of the struggle for human freedom. Although the term "materialism" is sometimes used today in philosophical discussions in a much more restrictive way to refer to the crude proposition that all natural processes are attributable directly to matter—and is in this way distinguished from naturalism, which is seen as attributing all natural processes simply to natural causes—we shall use the term materialism here in its classic sense, in which it is indistinguishable from naturalism. In this view, the defining trait of materialism from antiquity to the present has not been the forced adherence to a limited, metaphysical notion of "matter" as the all-encompassing reality (though the notion of the atom was essential to materialism from antiquity on and the concept of matter remains crucial to all science), but rather its opposition to all teleological explanations, i.e., final causes (whether God or Logos).

In its most general sense, then, materialism claims that the origins and development of whatever exists is dependent on natural processes and "matter," that is, a level of physical reality that is independent of and prior to thought. Materialism understood in this way can also be identified with the realist ontology characteristic of scientific realism.[18] As Bertrand Russell observed early in the twentieth century, materialism "has persisted down to our own time," from its beginnings in Greek philosophy, "in spite of the fact that very few eminent philosophers have advocated it. It has been associated with many scientific advances, and has seemed, in certain epochs, almost synonymous with a scientific outlook."[19]

Materialism's opposition to Platonic forms, absolute ideas, idealism, God, spirit, final causes, supernatural phenomena, miracles, Heaven, Hell, etc. in explaining the world, has made it the enemy of all forms of philosophical idealism, which invariably reach back to God, spirit, Logos. This essential aspect of the materialist world-view was well captured in the late nineteenth century by Frederick Engels, founder along with Karl Marx of historical materialism, who wrote in his *Ludwig Feuerbach and the Outcome of Classical German Philosophy*:

> "Did god create the world [the universe] or has the world been in exis-tence eternally?" The answers which the philosophers gave to this ques-tion split them into two great camps. Those who asserted the primacy of spirit to nature, and, therefore, in the last instance, assumed world cre-ation in some form or other—(and among philosophers, Hegel, for exam-ple, this creation often becomes still more intricate and impossible than in Christianity)—comprised the camp of idealism. The others, who regarded nature as primary, belong to the various schools of materialism. These two expressions, idealism and materialism, primarily signify noth-ing more than this; and here also they are not used in any other sense.[20]

Materialism is often treated by its critics, including intelligent design proponents, as inherently "reductionist," unable to account for the complexity of observable phenomena. It is true that there are mechanistic and reductive forms of materialism (perhaps better referred to as mechanism)—evident today in sociobiology, evolu-tionary psychology, and genetic determinism. However, a material-ist dialectic is essential, in our view, to a developed materialist out-look, if it is not to fall into the crude, reductionist perspective that the part determines the whole. Hence, we agree with biologists Richard Levins and Richard Lewontin that "in the dialectical world the logical dialectical relation between part and whole is [to be] taken seriously. Part *makes* whole, and whole *makes* part. . . . Organisms are both the subjects and the objects of evolution. They both make and are made by the environment and are thus actors in their own evolutionary history."[21] Since antiquity, materialist-dialectical thinkers have denied reductionism, mechanism, and determinism, along with teleology and religion. Forced to choose between one and the other, Epicurus argued that it would be better to believe slavishly in the interventionist gods of the multitude, than submit to determinism and abandon the possibility of human freedom. Fortunately, he added, it is possible to break the bonds of fate with respect to both teleology and determinism.[22]

Scientists who operate in a society such as the United States today, in which a large part of the population still believes to some extent in supernatural causes and some form of creation, often feel compelled to deny (lest they appear irreligious) the ontological bases of materialism/naturalism within science. Such propositions are often treated as a kind of "metascience," which is said to have little to do with science itself, defined purely in terms of its method. Hence, officially scientific materialism is presented merely as *methodological* naturalism/materialism. The U.S. National Academy of Sciences, for example, has expressly taken the position that science is a form of knowledge that *methodologically* assumes that nature can be explained simply in terms of natural processes, *without any concomitant view that this encompasses all realms of reality*, in a larger, ontological (metaphysical) sense.[23] But in that regard scientists are often disingenuous, since they are far more likely than the general population to adopt a materialist/naturalist ontology and to deny creation in any sense whatsoever. As Lewontin and Levins have written, "Creationists quite accurately identify the ideological content of science, which is secular humanism, against the liberal formula that science is the neutral opposite of ideology."[24]

Probably the most ambitious attempt in recent years on the part of a leading scientific figure (and by a materialist influenced by Marxism) to make room simultaneously for both religion and science was Stephen Jay Gould's *Rocks of Ages*. Gould, following an admirable instinct to make peace, famously presented his NOMA principle (Non-Overlapping Magisteria) to indicate that science and religion had their own independent domains (magisteria) and should therefore not be in conflict. Science addresses questions about the material world, whereas religion tackles questions of human meaning. Gould argued that there was no need for science and religion to be at odds, as long as each stuck to its own legitimate domain. He went on to note that historically, science and religion have not been in conflict as often or as intensely as is often assumed. For example, he noted that many of the historical conflicts over science in the West occurred *within* the Church, among people who shared religious faith. Furthermore, Gould contended that the methodological practices of science, although they require a "bench-top materialism," do not necessitate the renunciation of religious faith, as long as such faith does not impose strictures on the natural world. Science deals with the world of fact, religion with the world of faith and ethics.

However, despite the merits of Gould's argument, his proposed peace treaty was unlikely to work, since those of a devout religious persuasion are not disposed to concede the world of nature and "fact" to science. Intelligent design proponents frequently refer to science as "imperialistic," and complain that it seeks to take over more and more of God's domain. Gould's own solution involved giving the "morality of morals" (questions about what morals we *ought to* have) to the magisterium of religion (and the humanities) and the "anthropology of morals" (questions about what moral systems we *do* have) to the magisterium of science.[25] Gould is certainly correct that the anthropology of morals cannot lead to the morality of morals—as Hume famously noted there is no logical way to go from the *is* to the *ought*. Conversely, it is doubtful whether there is a foundationalist *ought* (as commonly presupposed by religion), that can tell us what moral values *should* be. Therefore, it is left to humans to construct their own morals.[26]

The "morality of morals," insofar as this involves transcendent moral principles, is from a thoroughgoing materialist standpoint stripped of historical meaning. Morality is something that humans must struggle with, individually and collectively and under changing conditions. Ethical conventions are to be viewed not in foundationalist, but in social-contractual terms.[27] Questions about what morals we ought to have cannot be answered in a factual manner, which leaves all factual questions about morality in the domain of science. Consequently, in the end, as intelligent design proponents rightly recognize, Gould's NOMA principle gives little or nothing to religion.

Further, although the naturalist methodology of science can and has been practiced by numerous religious believers from Newton to some scientists of today, the fundamental materialist philosophical position of science (which can hardly be relegated to mere "metascience") is inalterably opposed to any and all invocations of the supernatural, including the notion of divine providence. As Gould himself pointed out, "Darwin's intellectual radicalism emerges most clearly in the nature of natural selection as a materialist theory about a history of life without sensible purpose or necessary progress." Here the conflict with traditional religion was quite stark. As Alan Sokal has recently written, "The modern scientific worldview, if one is to be honest about it, leads naturally to atheism—or at the very least to an innocuous deism or pan-spiritualism that is incompatible with the tenets of all the traditional religions—but few scientists dare to say so publicly."[28]

Although we agree with Gould that in principle religion and science can coexist with a mutual non-aggression pact (Epicurus did not abolish the gods but simply expelled them from all relation to nature), the fact remains that the tension between religion and science is a deep-seated product of the alienated nature of our entire society. Both religious alienation and the alienation of science (which is relegated to serving the "gods of production and profit," in Rachel Carson's apt phrase) are necessary, if contradictory, elements in a structure of power. This is best understood from the standpoint of the critique offered by Marx, presented in chapter 5.[29]

If the issue of science versus religion allows for some degree of compromise at least at a practical level, the conflict between science and today's intelligent design creationism is absolute, precisely because the latter seeks to account for nature supernaturally. The demarcation problem within the philosophy and sociology of science has given rise to endless debates about the criteria distinguishing science from non-science. It is therefore common among anti-realist sociologists of science today to argue that there are no universal rules allowing for such a demarcation, which are determined rather by scientific consensus. Yet, to say that intelligent design is absolutely outside of and opposed to science, as we do here, is not to confront the difficult demarcation problem, since intelligent design's objective has never been to provide new scientific explanations. Rather, it seeks to make arguments to establish the limits of science (what might be called the "God limits"). Thus intelligent design proponents invariably point to gaps for which they say science has no explanation, and can have no explanation, and treat that as final, and indeed irrefutable, evidence of a supernatural Designer. Those who engage in science, in contrast, invariably seek to explore phenomena for which "science has no [adequate] explanation—yet."[30]

It is undeniable that science in all of its manifestations is currently under siege from the forces of religious irrationalism in the form of the intelligent design movement. The armistice between religion and science has been broken. There is consequently very little room today for intellectual compromise between materialist science and religion—and none between materialism and design. Indeed, Gould's own NOMA principle is harshly criticized by intelligent design proponents who call it a form of "apartheid" meant to segregate God from the world.[31]

Therefore, we frankly acknowledge that the conflict between religion and science is a permanent feature of our present-day capitalist society. The further advance of a humanistic science is necessarily a revolutionary project. At the same time, intelligent design is to be regarded as a reactionary movement directed against past gains. As Thomas Henry Huxley wrote, triumphantly, shortly after the publication of the *Origin of Species*, "Extinguished theologians lie about the cradle of every science as the strangled snakes beside that of Hercules, and history records that whenever science and dogmatism have been fairly opposed, the latter has been forced to retire from the lists, bleeding and crushed, if not annihilated; scotched if not slain."[32] Intelligent design is, in this sense, a counterrevolution against science.

Epicurus, Darwin, Marx, and Freud

In its latest resurrection with a vengeance, creationism in the form of intelligent design seeks not so much to triumph over materialism in public schools and other institutions as to burn it on the cross. Intelligent design proponents see the argument from design as part of a larger crusade against materialism that traces the problem not to Darwin but to Epicurus in antiquity. Epicurus is regarded as the archetype of materialism and the greatest single enemy of creationism. Hence, the refutation of Darwin is seen as necessary but not as the final or sufficient goal in a much larger inquisition. Indeed, intelligent design criticisms embrace the entire materialist tradition extending from Epicurus, viewed as a kind of Antichrist, to the unholy trinity of Darwin, Marx, and Freud in modern times.

According to William Dembski, senior fellow of the Discovery Institute's Center for Science and Culture and one of intelligent design's leading proponents, "All roads lead to Epicurus and the train of thought he set in motion."[33] Similarly, Benjamin Wiker, also a senior fellow of the Center for Science and Culture and its leading social philosopher, states: "Epicurean materialism was defined against every account of nature leading to an intelligent designer, and so it also always set itself against any religion which asserted that the universe was created and controlled by divine power."[34] Understood in this way, Wiker contends:

> Darwinism is part of a much larger theoretical and moral worldview, that of materialism . . . that . . . can be traced all the way back to the ancient Greek Epicurus. . . . As it turns out, our present moral state of affairs,

morbid as it is, is the result of having accepted the entire materialist package, of which Darwinism was an essential part. This larger materialist package supports all kinds of things which are morally repugnant to Christians, not only . . . Social Darwinism and eugenics, but also sexual libertinism, abortion, infanticide, euthanasia, cloning, and so on. . . . [W]e find out by reading Epicurus and Lucretius that materialism was designed to destroy all religion. When Christianity arose on the scene, not too long after Lucretius wrote his Epicurean materialist epic poem, it showed itself to be immediately antagonistic to Epicurean materialism. This fundamental antagonism can be traced historically over the next millennium and a half.[35]

Intelligent design proponents thus routinely present Darwin, Marx, and Freud as the modern representatives of a long tradition of materialism-humanism with its roots in Epicurus. Because of this, Epicurus, Darwin, Marx, and Freud are the four main targets of the intelligent design movement. Not content to criticize the role of materialism in the natural sciences, the intelligent design argument is extended to the social sciences, too. This serves to bring to the fore the goals of the intelligent design movement, which are as much a part of a struggle over human freedom as a struggle over nature, over social science as much as natural science.

The truth is that each of the leading thinkers of modern social science, along with many of the great thinkers in natural science, had to return to the critique of design and teleology and thus to the materialist roots of science as a prerequisite to the development of their views. Yet, this long critique of design, which was so integral to the development of science in all its forms, is little understood today, leaving those who wish to oppose the argument by design ill-equipped for the current struggle. Moreover, the relation of the materialism/design debate to the development of social science in particular is almost entirely overlooked outside the work of design proponents themselves.

It is widely recognized that Darwin needed to challenge the prevailing religious worldview in order to establish a foundation for rational inquiry into the processes of the natural world. This led him, as Gould pointed out, to apply "a consistent philosophy of materialism in his interpretation of nature. Matter is the ground of all existence; mind, spirit, and God as well, are just words that express the wondrous results of neuronal complexity."[36] But it is much less well known that a similar challenge lay at the foundation of the social sciences. In particular, Marx's effort to found a science of society paralleled those of Darwin (and Epicurus before him) and led him to dismantle the religious dogma of his day so as to build a materialist philosophy that enabled social analyses that were free of irrationalism. Likewise, Freud found it necessary to challenge theistic premises in his efforts to create a science of the mind. Thus, rather than needing to develop a new defense of social science against the critiques of intelligent design creationists, the social sciences have such a defense already prepared by Marx and (somewhat more problematically) by Freud as well. Despite their importance and force, Marx's challenges to intelligent design are generally neglected. Further, most social scientists today are largely unaware both that intelligent design seeks to challenge the foundations of social science and that a materialist defense of social science against these critiques already exists. Thus, our goal is to revive this neglected defense and link it to the long line of materialist inquiry going back to Epicurus so as to highlight the deep connections between the natural and social sciences, while providing a bulwark against the forces of irrationalism that seek to undermine both.

Not the least of the ironies surrounding the intelligent design movement's attack on Epicurean materialism as the classical critique of intelligent design and the forerunner in this respect of Darwin, Marx, and Freud, is that Marx, who wrote his doctoral thesis on Epicurus, has long been recognized by Epicurean scholars as one of the most penetrating nineteenth-century analysts of Epicurean materialism.[37]

At the base of Epicurus's materialism was a conception of contingency in nature and in the human world. "From the very outset," he wrote in *On Nature*, "we always have seeds directing us some towards these, some towards those, some towards these *and* those, actions and thoughts and characters, in greater and smaller numbers. Consequently that which we develop—characteristics of this or that kind—is at first absolutely up to us."[38] It is this conception of human freedom, based on material conditions and a relationship with the earth, proceeding *without the aid of the gods*, that constitutes the main threat to intelligent design creationism, a threat embodied in modern times by the work of Darwin, Marx, and Freud, in particular.

2. The Wedge Strategy

Today's intelligent design proponents, as one of their leading critics, Eugenie Scott, executive director of the National Center for Science Education, has stated, are divided as to the nature of design activity itself, which could take such varied forms as "front-loading all outcomes at the big bang, episodic intervention of the progressive creationism form, or other, less well-articulated possibilities." However, theistic evolution—the notion that God created the physical universe and has kept "his" hands off it ever since, allowing it to evolve via natural laws (except for the production of the human soul)—is "ruled out."[1] Hence, intelligent design proponents are not simply believers in God's creation in the very broadest sense, which holds to the notion that once God created the universe natural processes took over—a view that is consistent with theistic evolution. Rather they also give numerous indications of believing in what is known as "special creationism," in which it is held that the world was created in essentially the same form in which it exists today. They also typically assert that it can be inductively demonstrated and inferred that an intelligent designer must have had an active hand in the ongoing formation of the world.

Crucial to the intelligent design argument, as *Darwin on Trial* author Phillip E. Johnson, a legal scholar and program advisor to the Discovery Institute's Center for Science and Culture, says, is the notion that "God has influenced the creation on a regular basis."[2] Johnson and other leading intelligent design spokespersons have equated intelligent design with the term "mere creation" (a play on the title of C. S. Lewis's famous work of Christian apologetics, *Mere Christianity*). The intent is to establish a unified Christian anti-evolution movement, bringing together all creationists, including biblical literalists such as young-earth creationists, i.e., those who believe that the earth is no older than what is suggested in the Bible, as well as earlier versions of "creation science."[3]

This new intelligent design creationism is instigating a renewed war between religion and science that is potentially more virulent than any that occurred in the United States in the twentieth century. Intelligent design proponents defy the scientific consensus and draw for their support on the vast popular appeal of creationist views. A Gallup poll in November 2004 indicated that 45 percent of the U.S. population believes that human beings were created by God in their present form sometime in the last 10,000 years. Another 38 percent believes that human beings evolved with God's guidance. Only 13 percent are convinced that God had no part in the process. It is this widespread belief in creationist doctrines, along with widespread ignorance of evolutionary theory, that has allowed intelligent design to constitute itself as a popular educational movement.[4]

The new creationism is not simply a movement on the margins of society. It has received strong support from influential conservative political actors and opinion makers such as Ronald Reagan, Edwin Meese, George Gilder, William F. Buckley, Tom DeLay, John McCain, and George W. Bush. In 1980, when running for president, Ronald Reagan argued against the theory of evolution, saying that "recent discoveries down through the years have pointed up great flaws in it." Asked, while speaking to a fundamentalist religious coalition in Dallas, about teaching evolution in public schools, Reagan responded:

Well, it is a theory, it is a scientific theory only, and it has in recent years
been challenged in the world of science and is not yet believed in the scientific community to be as infallible as it was once believed. But if it was
going to be taught in the schools, then I think that also the biblical theory of creation, which is not a theory but the biblical story of creation,
should also be taught.[5]

The Reagan administration and the fundamentalist political-
religious movement with which it was closely associated clearly
provided the incubus for the new wave of intelligent design creationism that was to emerge in the 1990s. The Seattle-based
Discovery Institute was founded in 1990 by Bruce Chapman, who
had served as deputy assistant to the president, director of the
White House Office of Planning and Evaluation, and director of
the U.S. Census Bureau in the Reagan White House. He was aided
in Discovery's creation by his longtime friend George Gilder, a
leading proponent of supply-side economics (Reaganomics).
Chapman has remained the Discovery Institute's president since
its inception, Gilder is a senior fellow. Edwin Meese, U.S. attorney
general in the Reagan administration, is an adjunct fellow.

The Discovery Institute started out as part of the Hudson
Institute, a conservative think tank, and then spun off as an independent entity. In the late 1990s it emerged as the main institution-
al force behind the intelligent design movement through its Center
for the Renewal of Science and Culture (later renamed the Center
for Science and Culture), established in 1996 with the backing of
multimillionaire financier and evangelical Christian, Howard
Ahmanson Jr. The founding of this new institutional-intellectual
base for creationism provided the impetus for an upsurge in attacks
on the teaching of evolution in the public schools throughout the
nation.[6]

The influence of the new creationist movement can also be seen
in its growing political influence. In 1999 U.S. House Majority
Leader Tom DeLay used the Columbine school shootings as an

occasion to blame evolution for a whole list of social ills in the nation, including school shootings. "Our school systems," he opined, "teach the children that they are nothing but glorified apes who have evolutionized out of some primordial soup of mud." In 2001 U.S. senator Rick Santorum of Pennsylvania proposed an amendment to the No Child Left Behind legislation that would have described biological evolution as "controversial" (the amendment was not included, however, in the final No Child Left Behind Act). President George W. Bush has several times indicated support for teaching intelligent design in public schools. *National Review* editor at large William F. Buckley added his support to the intelligent design crusade in February 2007, insisting that due to arguments of Johnson and other intelligent design proponents evolution had lost "its title to exclusivity." Buckley was responding specifically to a controversy that arose in 2007 when presidential candidate John McCain delivered a speech at a luncheon hosted by the Discovery Institute. McCain had given his support two years earlier to teaching intelligent design in the public schools (though he left it open whether this should be in science classes).[7] In 2004 the Discovery Institute opened a Washington, D.C., office to promote its political, education, and cultural agenda.

The Origins of the Wedge

The attack on Darwinism, although viewed as crucial in its own right, is seen by intelligent design proponents as simply the thin end of a wedge, which, when hammered in, thickens into a full-fledged attempt to replace materialist philosophy, politics, and culture with fundamentalist Christian versions of the same.[8] As intelligent design proponent William A. Dembski succinctly expressed it, "Naturalism is the disease. Intelligent design is the cure."[9]

This was made clear by Johnson, writing in 2000 in *The Wedge of Truth: Splitting the Foundations of Naturalism*:

The Wedge of my title is an informal movement of like-minded thinkers
in which I have taken a leading role. Our strategy is to drive the thin edge
of our Wedge into the cracks in the log of naturalism by bringing long-
neglected questions to the surface and introducing them into public
debate. Of course the initial penetration is not the whole story because
the Wedge can only split the log only if it thickens as it penetrates.[10]

In Johnson's metaphor, the "thin edge" of the wedge was the
attack on Darwinian science, while the thick end represented a
much more ambitious strategy of overturning the entire materialist conception of the world of contemporary society. This was
articulated at greater length in the now notorious 1999 *Wedge
Strategy* document (more commonly known as the *Wedge
Document*)—a ten-page or so plan for the development of the intelligent design movement crafted by the Discovery Institute's
Center for Renewal of Science and Culture. According to the
opening paragraphs of the *Wedge Document* it was not natural and
physical science as much as materialist social science and culture
that were ultimately at issue:

The proposition that human beings are created in the image of God is
one of the bedrock principles on which Western civilization was built. Yet
little over a century ago, this cardinal idea came under wholesale attack
by intellectuals drawing on the discoveries of modern science.
Debunking the traditional conception of both God and man, thinkers
such as Charles Darwin, Karl Marx, and Sigmund Freud portrayed
humans not as moral and spiritual beings, but as animals or machines
who inhabited a universe ruled by purely impersonal forces and whose
behavior and very thoughts were dictated by the unbending forces of
biology, chemistry, and environment. This materialist conception of reality infected virtually every area of our culture, from politics and economics to literature and art.

The cultural consequences of this triumph of materialism were dev-
astating. Materialists denied the existence of objective moral standards,
claiming the environment dictates our behaviors and beliefs. Such moral

relativism was uncritically adopted by much of the social sciences, and it still undergirds much of modern economics, political science, psychology, and sociology. . . .

[M]aterialism spawned a virulent strain of utopianism. Thinking they could engineer the perfect society through the application of scientific knowledge, materialist reformers advocated coercive government programs that falsely promised to create heaven on earth.

Discovery Institute's Center for the Renewal of Science and Culture seeks nothing less than the overthrow of materialism and its cultural legacies.[11]

By making their ultimate objective (the thick end of the wedge) the destruction of materialist social science and culture, the intelligent design proponents showed that the attack on Darwinism was conceived as part of a larger grand strategy. The *Wedge Document* outlined this strategy in terms of five-year and twenty-year projects and goals. Its projects were divided into three phases: "Phase I: Scientific Research, Writing, and Publication." "Phase II: Publicity and Opinion-Making." "Phase III: Cultural Confrontation and Renewal." Its "Twenty Year Goals" included not only seeing design theory enter into the physical sciences, but also "psychology, ethics, politics, theology and philosophy in the humanities." Included in the *Wedge Document* was a plan to alter contemporary views on "sexuality, abortion and belief in God." Indeed, the ultimate objective of the wedge strategy was "to see design theory permeate our religious, cultural, moral and political life"—or, as Phillip E. Johnson put it at one time, to replace the "ruling philosophy of modern culture," i.e., naturalism, with the new ruling philosophy of intelligent design.[12]

All of the leading proponents of intelligent design, including Phillip E. Johnson, William A. Dembski, Michael Behe, Stephen Meyer, Jay Wesley Richards, John G. West, Benjamin Wiker, Jonathan Witt, Jonathan Wells, Nancy Pearcey, and Charles Thaxton, signed on to this strategy to which subsequent works by intelligent design proponents have closely conformed. So important is this strategy to their activities that intelligent design propo-

THE APOSTLES OF INTELLIGENT DESIGN

President, Discovery Institute
Bruce Chapman

Discovery Institute's Center for Science and Culture
Program Advisor
Phillip E. Johnson

Program Director
Stephen C. Meyer

Associate Director
John G. West

Senior Fellows
Michael J. Behe
David Berlinski
Paul Chien
William A. Dembski
David DeWolf
Guillermo Gonzalez
Michael Newton Keas
Jay W. Richards
Jonathan Wells
Benjamin Wiker
Jonathan Witt

Fellows
(incomplete list)
Dean Kenyon
Robert C. Koons
Nancy Pearcey
Charles Thaxton
Richard Weikart

Source: Discovery Institute Web site, http://www.discovery.org

nents have sometimes called their movement "the Wedge." For example, Johnson refers to "the Wedge as an intellectual movement," and in promoting intelligent design proponents he sometimes uses the term "Wedge members."[13] Almost all of the leading members of the Discovery Institute's wedge movement are conservative and fundamentalist Christians—mainly evangelical Protestants and Catholics—although one, Jonathan Wells, is a member of Reverend Sun Myung Moon's Unification Church.[14]

A "Top Secret" Document

The *Wedge Strategy Document* evolved gradually. A key event was the summer 1995 "Death of Materialism and the Renewal of Culture" conference, sponsored by Johnson, out of which the Center for the Renewal of Science and Culture arose. The opening paragraphs of an introduction by political scientist John G. West to a spring 1996 symposium in the journal *Intercollegiate Review* on "The Death of Materialism and Renewal of Culture"—evolving out of this conference—were virtually identical with the opening paragraphs of the 1999 *Wedge Document*, as quoted above. Thus it was West who first observed that in "debunking the traditional conceptions of both God and man, thinkers such as Karl Marx, Charles Darwin, and Sigmund Freud portrayed human beings not as eternal and accountable beings, but as animals or machines" (though in West's original article, as opposed to the *Wedge Document*, Marx came before Darwin). From this it is clear that West was a major author of the *Wedge Document*.[15]

The year after the "Death of Materialism" conference, another landmark gathering was held at Biola University called the "Mere Creation" conference. In a "Postscript" to the book *Mere Creation*, bringing together the lectures from this conference, Discovery Institute president Bruce Chapman revealed the objectives of the new movement, stating:

Materialism is not limited in its implications to natural science. Materialism is a way of understanding day-to-day existence and responding to it. Materialism has influenced public standards and policies on morals, law and criminology, education, medicine, psychology, race relations, the environment, and many other areas.

It can be argued that materialism is a major source of the demoralization of the twentieth century. Materialism's explicit denial not just of design but also of the possibility of scientific evidence for design has done untold damage to the normative legacy of Judaeo-Christian ethics. A world without design is a world without inherent meaning. In such a world, to quote Yeats, "things fall apart; the center cannot hold."

Materialism not only prevails in the natural sciences but has also been adopted by such soft sciences as sociology and psychology. . . . [I]f the materialist interpretation of science is wrong, so surely is its application and misapplication in public policy and culture.[16]

By 1996 the main components of the wedge strategy were therefore already in place. Nevertheless, the actual circumstances surrounding the origins and dissemination of the *Wedge Document* remain shrouded in mystery. According to Roger Downey, writing for the *Seattle Weekly* in February 2006 on "Discovery's Creation":

The story begins, so far as the world at large is concerned, on a late January day seven years ago, in a mail room in a downtown Seattle office of an international human-resources firm. The mail room was also the copy center, and a part-time employee named Matt Duss was handed a document to copy. It was not at all the kind of desperately dull personnel-processing document Duss was used to feeding through the machine. For one thing, it bore the rubber-stamped warnings "TOP SECRET" and "NOT FOR DISTRIBUTION." Its cover bore an ominous pyramidal diagram superimposed on a fuzzy reproduction of Michelangelo's Sistine Chapel rendition of God the Father zapping life into Adam, all under a mysterious title: *The Wedge*.

Curious, Duff rifled through the 10 or so pages, eyebrows rising ever higher, then proceeded to execute his commission while reserving a copy of the treatise for himself. Within a week, he had shared his find with a friend who shared his interest in questions of evolution, ideology, and the propagation of ideas. Unlike Duss, Tim Rhodes was technically savvy, and it took him little time to scan the document and post it to the World Wide Web, where it first appeared on Feb. 5, 1999.[17]

In this way the Center for Renewal of Science and Culture's *Wedge Document* was made public on the Internet through the intervention of a third party, despite the Discovery Institute's intention to keep the document itself "top secret." (A *New York Times* article suggested that it was originally intended as a "fund-raising pitch.") It became known to critics of intelligent design in March 1999 and since then has been widely circulated and continuously available online. Although the Discovery Institute itself never posted the *Wedge Document* in its entirety on its Web site, it did post the opening paragraphs, originally written by West, on the site. The *Wedge Document*'s authenticity was acknowledged in 1999 by Jay Wesley Richards, then program director of the Center for the Renewal of Science and Culture, who admitted that it was "an older, summary overview of the 'Wedge' program." Much of the program was to be found in essentially the same words on the Center's Web site, though not the entire *Wedge Document*.[18]

What made the Discovery Institute's 1999 *Wedge Document* so important (and so "top secret" initially) was that it laid out the long-term strategy of the intelligent design movement, including its five-year and twenty-year objectives, and its clear suggestion that the ultimate goals of the wedge strategy were theological, political, cultural, and moral, rather than "scientific." As critic Eugenie Scott describes the wedge strategy, "The second focus of ID [intelligent design] is 'cultural renewal,' a term its proponents use to describe the movement's efforts to replace the alleged philosophical materialism of American society with a theistic (especially Christian) religious orientation."[19] Attacks on materialist science are therefore viewed by intelligent design proponents as leading to an attack on materialism more generally, just as their advocacy of design ("mere creation," as Johnson calls it) is seen as a means of ultimately extending God's dominion over the moral world (displacing materialist philosophy and culture).

C. S. Lewis, Christian Apologetics, and Mere Creation

The intelligent design movement begins and ends with an attack on materialism/naturalism, first in relation to science, and then at the thick end of the wedge, in relation to the entire culture. The inspiration for this strategy can be traced primarily—as the intelligent design proponents themselves make clear—to the Christian apologetics of C. S. Lewis. The Discovery Institute was originally founded on the basis of Lewis's apologetics and one of its programs is "C. S. Lewis and Public Life." The associate director of the Center for Science and Culture, John G. West, is a Lewis scholar, co-editor of *The C. S. Lewis Readers' Encyclopedia*. West's 1996 lecture "C. S. Lewis and the Materialist Menace" to be found on the Discovery Institute Web site represents one of the key statements of the intelligent design philosophy. It sees Lewis's offensive on the materialism associated with the work of Charles Darwin, Karl Marx, and Sigmund Freud as the first step in the renewal of Christian apologetics. Lewis, we are told, "was calling us back to a teleological view of the universe of the sort offered by Aristotle." Materialism for Lewis, as explained by West, gave rise to the moral "relativism" that "was uncritically adopted by much of the social sciences, and . . . still undergirds much of modern economics, political science, psychology, and sociology."

In his introduction to the symposium on the "Death of Materialism and Renewal of Culture," West, in line with Lewis, singled out sociology for its denial of any "universal standard" of morality. For West it is significant that Lewis made Mark Studdock, a young sociologist, the central focus of his novel *That Hideous Strength*, the last installment (preceded by *Out of the Silent Planet* and *Perelandra*) in his well-known space trilogy. Studdock's soul was sought by the forces of evil associated with the technocratic

National Institution for Coordinated Experiments (NICE) and thus became the subject of a conflict between the powers of light and the powers of darkness. At one point in the novel, in an exchange with a chemist, Studdock, who is still entrapped in his materialist delusions, refers to "sciences like Sociology." The chemist responds: "There *are* no sciences like Sociology." This, according to West, reflected Lewis's strong rejection of social sciences dominated by materialism, of which sociology was the leading form. In the end, Studdock is forced to confront his own materialism and moral relativism, recognizing his personal responsibility (sins), thus opening the way to his redemption. (In the novel's closing pages he is saved through the intervention of a newly arisen Merlin together with a Christian enclave, and helped out by heavenly powers. Christianity thus triumphs in this one struggle; in a world dominated by but not completely won over to evil.)

In a retrospective review of Lewis's *That Hideous Strength* written in 2000, Johnson depicted it as an attack on "materialist philosophy" and on "what eventually happens when people make technology their lord instead of putting their faith and love to the service of the true Creator."[20]

C. S. Lewis's legacy of anti-materialism and Christian apologetics has been continually drawn upon in the work of intelligent design proponents Dembski, Richards, Wiker, and Witt, along with Johnson and West. Lewis's open apologetics helped to inspire the 1998 intelligent design volume, *Unapologetic Apologetics: Meeting the Challenges of Theological Studies*, edited by Dembski and Richards. As Richards wrote in this collection:

> The naturalist just means everything when he says the word *nature*. C. S. Lewis puts it this way: "Just because the Naturalist thinks that nothing but Nature exists, the word Nature means to him merely 'everything' or 'the whole show' or 'whatever there is.' And if that is what we mean by Nature, then of course nothing else exists." The Christian or other theist will inevitably deny this claim, insisting that there is a greater self-subsist-

ing Reality, namely God, who himself created and sustains nature. Nature as such is a dependent, and not the ultimate, reality.[21]

For Dembski and Richards the main *religious* goal is to question those versions of Christian theology that have succumbed to naturalism by moving away from the concept of design. Such "naturalism," Richards exclaimed, "contradicts Christian belief." Indeed, "as C. S. Lewis puts it, 'a naturalistic Christianity leaves out all that is specifically Christian.'" Like Lewis, Richards argued straightforwardly for a "supernaturalist" rather than "naturalist" position.

No doubt inspired by the acceptance of "perplexity" in the face of the factual disproof of faith, which Lewis had openly displayed in his "Christian Apolgetics," Dembski argued in *Unapologetic Apologetics* for the error-free nature of scripture. Although biblical scripture may seem to have been contradicted by the facts as revealed by materialism, those who adopt the principle of "God's-eye point of view," Dembski claimed, will on such occasions opt for "perplexity." The latter is based on the recognition of our own limited understanding, and thus prevents us from wrongly assuming that scripture has been proven erroneous. "The choice then is up to us, which perspective we are going to trust," Dembski writes, "ours or God's?" As his intelligent design colleague West expressed it, "The Bible is infallible; but its interpreters are not."[22]

C. S. Lewis's Christian apologetics were based on a denial of materialism from its inception in antiquity to the present. A "cosmic futility," he suggested in his essay "*De Futilitate*" in *The Seeing Eye*, could be traced back as far as Lucretius (basing his views on Epicurus). In his youth, as he recounted in his autobiographical *Surprised by Joy*, Lewis had studied Democritus, Epicurus, and Lucretius, and had developed an affinity for the Epicurean materialist view. He was to renounce this along with much else on becoming a devout Christian. Hence, a sense of profound conflict between materialism and creationism pervaded his thought.[23]

Although it is clear that Lewis's works on Christian apologetics as a whole, including such works as *Mere Christianity*, *The Abolition of Man*, and *The Problem of Pain*, deeply influenced the wedge movement, it was his 1947 book *Miracles*, his last major work on apologetics, that most directly foreshadowed today's intelligent design outlook. In *Miracles* Lewis drew a very sharp distinction from the start between naturalism (or materialism) and supernaturalism. "Some people," he wrote, "believe that nothing exists except Nature; I call these people *Naturalists*. Others think that, besides Nature, there exists something else: I call them *Supernaturalists*." Supernaturalists, among whom Lewis counted himself, believe that there is "some self-existent thing," which "we call God" from which all else is "derived." "What Naturalism cannot accept," he stated, "is the idea of a God who stands outside of Nature and made it." Naturalism (or materialism) thus views nature as a "total system." Indeed, what Lewis objected to the most with regard to such a "total system" was that "there was no Designer." In his "Two Lectures" (1945), later reprinted in *God in the Dock*, Lewis adopted the classic argument from design, stating that just as a rocket has a designer, so must nature. It was therefore a great tragedy, he suggested, that naturalism/materialism was the dominant outlook of the modern world. "We all have Naturalism in our bones," he wrote, "and even conversion does not at once work the infection out of our system."[24]

In defiance of what he viewed as this hell-like naturalism, Lewis sought self-consciously to advance Christian apologetics. He opted for an alienated view of reality consisting of God (or the Trinity) and Nature. "God created Nature. This at once supplies a relation between them and gets rid of sheer 'otherness.'" Nature was in no way God's equal but "a *creature*, a created thing." Nature, it was true, normally operated by its own laws, but the supernatural might intervene in the form of "*selected* events" or "miracles." All miracles were merely subsidiary to the "grand miracle," the incarnation of Christ aimed at redemption of the world.[25]

Although rightly belonging to God's domain, nature, was rebellious and the world was a "fallen world." "Nature by dominating
spirit," Lewis wrote, "wrecks all spiritual activities." Consequently,
he often invoked a "state of war" between nature and the reason of
the heavenly powers. In this war, nature, however, was not the equal
of reason (Logos). "Nature can only raid Reason to kill; but Reason
can invade Nature to take prisoners and even to colonise."

Or as he put it somewhat differently elsewhere in the text:

> Our whole picture of Nature being "invaded" (as if by a foreign enemy)
> was wrong. When we actually examine one of these invasions it looks
> much more like the arrival of a king among his own subjects or a mahout
> visiting his own elephant. The elephant may run amuck, Nature may be
> rebellious. But from observing what happens when Nature obeys it is
> almost impossible not to conclude that it is her very "nature" to be sub-
> ject. All happens *as if* she had been designed for that very role.[26]

Still, Lewis constantly reverted to the notion of "invasion"—
even if, as he said, this could be more properly understood as a sov-
ereign reclaiming his rightful domain. He saw Christianity as ultimately engaged in a reconquest of the world, and hence a war
against naturalism/materialism (much like that propounded by the
wedge proponents): "There is no question in Christianity," he
wrote, "of arbitrary interferences just scattered about. It relates not
a series of disconnected raids on Nature but the various steps of a
strategically coherent invasion—an invasion which intends complete conquest and 'occupation.'" It is a war between good and evil.
"It is because Man is . . . a spiritual animal that he can become
either a son of God or a devil." In the end even death, which was
brought to the world by Satan and "man's fall," will be transcended, as foretold by Christ's "grand miracle."

Lewis made his antipathy to modern materialist science clear,
and saw the reconquest of nature by Christian notions of intelligent
design as an object. This required nothing less than the toppling of
contemporary science itself. As he hopefully declared in *Miracles*,

"We may be living nearer than we suppose to the end of the Scientific Age."[27]

It is no wonder, then, that C. S. Lewis has emerged as the patron saint of the intelligent design movement. The Discovery Institute Web page on "C. S. Lewis and Public Life" includes a quarterly journal, *The C. S. Lewis Legacy Online*, a section on "C. S. Lewis Writings in the Public Domain," and a series of articles on Lewis by West and other intelligent design proponents. No other thinker is accorded such elevated status within the intelligent design pantheon.

The Critique of Materialism versus the Critique of Intelligent Design

C. S. Lewis saw his critique of materialism and defense of intelligent design as part of a long history, stretching back to the ancient Greeks. He frequently aimed barbs at the Greek atomists, Spinoza, Hume, Marx, Darwin, Freud, J. B. S. Haldane, and others. Today's intelligent design proponents have likewise continually reasserted what Center for Science and Culture fellow Nancy Pearcey has called "The Long War Between Materialism and Christianity." Pearcey has described this debate as beginning with the early Christian thinkers who developed their views by "forcefully attacking Epicurean materialism." But since the roots of the controversy extend back to the pre-Christian Western world it might be more comprehensively called "The Long War Between Materialism and Design."[28]

The argument from design was first introduced by Socrates in ancient Athens in response to the development of materialist philosophy (particularly atomism). The critique of materialism by intelligent design that commenced with the Socratic philosophy resulted in a critique of intelligent design by materialism, beginning with Epicurus.[29] These debates continued for hundreds of years. Eventually, Christianity emerged on the side of intelligent design, presenting Epicurean materialism as its chief philosophical enemy. The Enlightenment resulted in a revival of materialism and the challenging of Christian teleology. In this long war, which has taken place over millennia, materialism has triumphed in the domain of nature and science, i.e., worldly philosophy. The very term "intelligent design" was introduced in its modern sense by Darwin in a powerful critique of teleological views, in which he argued that though nature may appear to be designed, there is in fact no intelligence behind it.

In the twentieth century creationism or intelligent design retreated into the realms of theology, Christian apologetics, idealist philosophy, and popular superstition. With the new intelligent design movement, it is now attempting once again to carry out, to quote Lewis, a "strategically coherent invasion—an invasion which intends complete conquest and 'occupation.'" The ultimate goal is not to dominate the natural sciences only or even primarily—but rather to dominate social science, culture, philosophy, morality, and public life. This is the political and religious end of the wedge.

To place all of this in historical context, it is necessary to explore the millennia-long intellectual struggle in this area. The critique of materialism represented by intelligent design has to be understood against the background of the critique of intelligent design by materialism—from antiquity to the present. What this demonstrates is no simple, straightforward story, but a long dialectical conflict.

3. Epicurus's Swerve

The phrase "intelligent design" first achieved public prominence in the nineteenth century in a discussion of the Epicurean materialist critique of the design argument. David Hume's 1748 *Enquiry Concerning Human Understanding* famously included a dialogue containing an imaginary defense by Epicurus before the Athenian masses justifying his rejection of the "religious hypothesis" of "intelligence and design" in nature.[1] But the oldest known use of the specific phrase "intelligent design" in its modern sense, as noted in the introduction, can be traced to none other than Darwin himself. Writing privately to John Herschel, one of the leading British scientists of the day, with regard to Herschel's 1861 article on "Physical Geography" for the *Encyclopedia Britannica* (which he had sent to Darwin), Darwin observed: "The point which you raise on intelligent Design has perplexed me beyond measure. . . . One cannot look at this Universe with all living productions & man without believing that all has been intelligently designed; yet when I look to each individual organism, I can see no evidence of this."[2]

Nevertheless, the most prominent early public use of the term "intelligent design" in its modern sense can be traced to the noted British physicist John Tyndall in his presidential address (often called the "Belfast Address") to the British Association for the

Advancement of Science in 1874.[3] Today Tyndall is best known as
the scientist who through his experiments first demonstrated that
water vapor, carbon dioxide, and methane acted as greenhouse
gases retaining solar heat on earth.[4] In his Belfast Address Tyndall
launched a defense of materialist science, speaking at length about
the role of Epicurus and his follower the Roman poet Lucretius (c.
99–55 BCE) in opposing teleological conceptions of the universe.[5]
In explaining how Lucretius's *De rerum natura* portrayed a universe based in atomism and governed by contingency and emergence, Tyndall stated:

> The mechanical shock of the atoms being, in his [Lucretius's] view, the
> all-sufficient cause of things, he combats the notion that the constitution
> of nature has been in any way determined by intelligent design. The
> interaction of the atoms throughout infinite time rendered all manner of
> combinations possible.... "If you will apprehend and keep in mind these
> things, Nature, free at once, and rid of her haughty lords, is seen to do all
> things spontaneously of herself, without the meddling of the gods."[6]

It is this materialist outlook, exemplified by ancient
Epicureanism, suggesting that nature can be understood as evolving
spontaneously into more complex, emergent combinations, developing by means of contingent occurrences, that most threatens creationist thinkers. Attached to this, in Epicurus's case, was a conception of social evolution and human freedom that rejected foundationalist ethics—that is, the gods as intelligent moral designers and
the existence of absolute moral principles independent of human
social contracts under changing conditions. Together these propositions made Epicurus and his followers in subsequent centuries the
great enemies of ancient teleology. For emerging Christianity no
greater philosophical threat existed than Epicurean materialism.

Origins of Critique of Intelligent Design

The intelligent design argument and teleological views more
broadly predate Christianity within Western civilization and can be

traced back to the ancient Greeks and Romans, including Socrates (469–399 BCE), Plato (c. 427–347 BCE), Aristotle (384–322 BCE), the Stoics (in Hellenistic and Roman times), Cicero (106–43 BCE), and Plutarch (c. 46–121); while the greatest ancient critic of teleology and intelligent design was Epicurus (341–270 BCE). The origin of the argument from design, i.e., that family of arguments that purports to provide proof of the existence of a creator god through evidence of design in nature, can be traced to Socrates, as depicted in Xenophon's (c. 428–354 BCE) *Memorabilia*. "Xenophon's Socrates," as classicist David Sedley has observed, was "a fundamentally anti-scientific creationist."[7] Socrates in the *Memorabilia* argued that human beings were uniquely favored by the gods, and exhibited the creator's design in their very being. In this view craftsmanship, as displayed, for example, in sculpture, provided an analogy for the supreme craftsmanship of a divine creator, who could not only produce forms of things but also give them life. Human beings, for Socrates, were thus clearly "products of design and not of chance," and demonstrated in their basic attributes the "intelligence" of the divine craftsman. The human eye (in its external features) was singled out as an example of this. As Socrates put it in his dialogue with Aristodemus:

> Then don't you think it was for their use that he who originally created men provided them with the various means of perception, such as eyes to see what is visible and ears to hear what is audible?. . . For example, because our eyes are delicate, they have been shuttered with eyelids which open when we have occasion to use them, and close in sleep; and to protect them from injury by the wind, eyelashes have been made to grow as a screen; and our foreheads have been fringed with eyebrows to prevent damage even from the sweat of the head.[8]

Such design, Socrates argued, could also be shown in other ways: in the granting of human beings, as distinguished from other animals, both intelligence (derived from cosmic intelligence) and non-seasonal sex; and in the way that a whole range of other species were created by divine agency specifically to serve human needs.[9]

Creationist views existed in Greek philosophy long before Socrates. But the need for a naturalistic defense of a creator god, i.e., an argument from design, only arose once the early Greek atomists Leucippus (fifth century BCE) and Democritus (c. 460–356 BCE) had introduced the notion that the material world and life within it emerged from the chance movement of atoms. The argument from design was thus from the beginning a response to materialist/atomistic physics. Socrates's intent in introducing the argument from design was to provide an idealist response to the materialist views propounded by the Greek atomists Leucippus and Democritus, whose work had thoroughly displaced the gods.[10]

In building on Socrates's thought, Plato did not directly advance the argument from design in the fashion of Xenophon's Socrates. He did, however, promote creationist ideas, and a creationist physics. For the ancient Greek philosophers matter was always a precondition of all else. Hence, Plato's Demiurge or divine craftsman in *Timaeus* did not create the world *ex nihilo* (out of nothing). Rather he relied on previously existing matter to generate order out of chaos. The Demiurge designed the world on the model of "the perfect intelligible Living Creature." In this way Plato's Demiurge constituted the greatest of all causes, and generated a world that was the finest of all possible worlds. In *The Laws* Plato urged that those who were impious, and attributed the world's coming into being to necessity and chance rather than design, be treated as criminals and imprisoned or even executed.[11]

Aristotle, although often considered to be the greatest teleological thinker in antiquity if not all time, was a step further removed from the argument from design. Everything in nature, according to his philosophy, was purposive. Yet, Aristotle's philosophy lacked a Demiurge as in Plato or a strict creationist argument. Still, a more distant deity, a kind of unmoved mover, is the supreme explanatory principle of Aristotle's philosophy. Everything in nature derives

its purposiveness in striving to emulate this divine "final cause." As
he wrote in *The Parts of Animals*: "Plainly...that cause is the first
which we call the final one. For this is the Reason, and the Reason
forms the starting point, alike in the works of art and in the works
of nature. . . . Now in the works of nature the good end and the final
cause is still more dominant than in works of art." For Aristotle the
purposiveness of nature, governed by a distant divine impulse or
final cause, was therefore invariably superior to the purposiveness
of human art.[12]

The most explicit proponents of the argument from design itself
in antiquity after Socrates were the Stoics, and Cicero, who identified with the Greek philosophical school of the Academy, i.e., was a
mild skeptic. The argument from design as articulated by Socrates in
Xenophon's *Memorabilia* was taken over directly, beginning in the
early third century BCE, by the Stoics, who considered this a basic
text.[13] In opposition to Epicurus and his followers, the Stoics propounded the teleological view of a providential god, the existence of
which could be ascertained from evidence of design in nature. The
resulting debate between Stoics and Epicureans over teleology and
design extended well into Roman times.

The most prominent Roman treatise to advance the argument
from design was Cicero's dialogue *The Nature of the Gods* (written
in 45 BCE), a work that included criticisms of Epicurean materialism. As Balbus, the Stoic, stated in Cicero's dialogue:

> When you follow from afar the course of a ship, upon the sea, you do not
> question that its movement is guided by a skilled intelligence. When you
> see a sundial or a water-clock, you see that it tells the time by design and
> not by chance. How then can you imagine that the universe as a whole is
> devoid of purpose and intelligence? . . . Our opponents however profess
> to be in doubt whether the universe...came into being by accident or by
> necessity or is the product of a divine intelligence. . . . The truth is that it
> [the universe] is controlled by a power and purpose which we can never
> imitate. When we see some example of a mechanism, such as a globe or
> a clock or some such device, do we doubt that it is the creation of a con-

scious intelligence? So when we see the movement of the heavenly bodies, the speed of their revolution, and the way in which they regularly run
their annual course, so that all that depends on them is preserved and
prospers, how can we doubt that these too are not only the works of reason but of a reason which is perfect and divine?[14]

For the Stoics, as portrayed in Cicero's dialogue, and no doubt
for Cicero as well, who was less of a skeptic in relation to religion,
the argument from design was seen as countering ancient materialism, particularly Epicureanism, with its proto-evolutionary views
and critique of intelligent design.[15] As A. A. Long, one of the foremost scholars of Epicureanism and Hellenistic philosophy in general, recently wrote in an essay titled, "Evolution vs. Intelligent Design
in Classical Antiquity": "The Epicureans even today are the unsung
heroes of ancient science if you are looking for significant anticipa-
tions of a modern rationalistic outlook. They are unsung mainly
because popular culture has preferred the theistic outlook of Plato
with its Biblical affinity. . . . What aligns them with our science is the
following set of methodologies and assumptions":

1. The starting point for understanding the world is rigorous empircism.
2. We have reason to think that everything we experience is ultimate-
    ly explicable by reference to physical facts and causes.
3. The building blocks of the world are uncreated and everlasting
    atomic particles incessantly in motion.
4. Science has no use for inherent purposiveness or mind in matter.
5. Apparent evidence for design in nature (for example, the complex-
    ity of organisms and organs) is due not to an invisible guiding hand
    but to the determinate ways that matter organizes itself according
    to strict causal laws.
6. Life and mind are not basic to the world, but emergent properties
    of particular types of atomic conglomerates.[16]

Not only did Epicurus and his followers attempt to advance
these propositions, but they did so not on the basis of faith but with
rational arguments, using a sophisticated method of scientific infer-

ence that was to influence later scientific thought.[17] The Epicureans, as Sedley writes, were "the ancient philosophical world's most ardent empiricists."[18]

Epicurus's philosophy was concerned above all with escaping the double trap (the bonds of fate) represented by the gods and mechanistic determinism. Adherence to the notion of the gods as prime movers in the world meant, in Epicurus's view, ascribing to an anti-scientific philosophy in describing the world. However, strict mechanistic determinism, while displacing the gods and allowing for a materialist science, denied human agency altogether.

Epicurus sought to escape both of these positions. Similar to modern scientists, he rejected explanations of the world based on final causes, particularly divine causation. As Lucretius put it, evoking a principle common to ancient Greek philosophy, "Nothing is ever created by divine power out of nothing."[19] Epicurus took this principle to its furthest point, rejecting all teleological positions, grounding the examination of the physical world in material (natural) explanations. Building on the earlier atomic theory of Democritus, Epicurus described the universe in terms of physical processes rather than final causes. In explaining the happenings of the world, he accepted the principle of multiple possible causes that only could be adjudicated by empirical investigation. He sought a general theory of causation, where a single correct explanation might not be possible given limitations in observing the exact phenomena (for instance, during his lifetime, with respect to solar and lunar eclipses). Instead, several alternative hypotheses were set up to account for any other conditions that might contribute to the relationship or event under investigation. It is from Epicurus that we get the phrase "awaits confirmation." In this, Epicurus "maintained his empirical principle that a scientific explanation must be consistent with, or not contradicted by, experience" and conform to a "general principle of determinism [material causation], without claiming to have knowledge of specific causes in all cases."[20]

Epicurus resisted a mere mechanical determinism without giving way to idealism. He claimed that the world was composed of atoms that continued to fall through the void, yet swerved, almost imperceptibly, as they fell, creating the element of chance and indeterminacy. These actions took place within and through material conditions; the swerve was both facilitated and limited by them. The existence of the swerve created added uncertainty in the course of life. Within a particular temporal context, at a specific point—as Lucretius explained—"accidents of matter . . . happen."[21] These accidents are the result of complex interactions, and the implication of these collisions is not known. Thus Epicurus saw contingency, due to the swerve of atoms, as an escape from the confines of gods and determinism. In fact, contingency is at the heart of change at every level and in every stage of life, and, as a result, novelty becomes part of history and life.[22]

Marx, who was arguably the most profound scholar of Epicureanism in the nineteenth century, understood Epicurus's attack on both mechanistic determinism and teleology as the basis of a doctrine of freedom that was extended into the social realm and human history. In Epicurus's Garden (as opposed to other ancient Athenian philosophical schools) women and slaves were admitted as equals to his society of friends, with some, such as Leontion (a *hetaera*), authoring noted philosophical treatises.[23] Jean-Paul Sartre, following Marx, wrote in his essay on "Materialism and Revolution": "The first man who made a deliberate attempt to rid men of their fears and bonds, the first man who tried to abolish slavery within his domain, Epicurus, was a materialist."[24]

Epicurus did not kill the gods. He simply separated them from the material world, banishing them to the pores—or *intermundia*, the spaces between the worlds—of the universe. Epicurus, Marx noted, saw the need for "the plastic gods of Greek art," but not gods as material actors.[25] As Alfred Lord Tennyson lamented in his 1868 poem *Lucretius*, the gods of Epicurus haunt:

The lucid interspace of world and world
Where never creeps a cloud, or moves a wind,
Nor ever falls the least white star of snow,
Nor ever lowest roll of thunder moans,
Nor sound of human sorrow mounts to mar
Their sacred everlasting calm![26]

Epicurus was concerned with combating the tendencies toward a state religion, based on a notion of astral gods, as presented by Plato in his *Laws*—a notion that was gaining influence in Hellenistic times. He expressly stated that the gods, though they exist, have no relation to the material universe, including the heavens themselves.[27] It was this classic version of what Stephen Jay Gould has called NOMA, or the notion of non-overlapping magisteria of science and religion—removing the gods from all connection to the material world and thus making it the magisterium of science—that most outraged Epicurus's critics for more than two thousand years.[28]

The rejection of all design or providence was to make Epicureanism anathema not only to Platonists like Plutarch, who in the second century wrote a polemic (*Against Colotes*) directed entirely against the first generation of Epicureans some four hundred years after the establishment of Epicurus's Garden, but also to all of the Church Fathers of early Christianity.[29] The latter condemned Epicurus for his rejection of both providence and the immortality of the soul. Epicurus was thus seen as the greatest ancient critic of design. As Howard Jones notes in *The Epicurean Tradition*, "From Athenagoras in the second century to Augustine in the fifth, we find repeated the familiar appeal to design as proof of the hand of an intelligent creator and the controller of the universe as against the random union and configuration of atoms posited by the Epicureans."[30] In his *City of God* St. Augustine attacked Epicurus for placing the criterion of truth in the "bodily senses," and went on to refer to "a world which bears a kind of silent testimony to the fact of its creation, and proclaims that its maker could have been none other than God." Later Thomas Aquinas in the thirteenth century supported Plato on the rule of eternal ideas and decried the views of "Democritus and the Epicureans," since they "denied that there is any providence" and "held that the world came about by chance." In direct opposition to Epicurus, Aquinas argued the case for design, stating that "inasmuch as natural things are without knowledge, there must be some pre-existing intelligence directing them to an end, like an archer giving a definite motion to an arrow to wing its way to that end. . . . So the world is governed through the providence of that intellect that gave to nature this order."[31]

In the fourteenth century, Dante's *Inferno*, Canto X (part of *The Divine Comedy*), reflecting similar views, consigned Epicurus and his followers to an eternity of torture in open coffins in the sixth circle of Hell.[32]

Despite Epicurus's frequent references to the gods and his defense of religious piety, his rejection of design and of the immortality of the soul generated continual charges of "atheistic materialism"—directed at him over the millennia. Epicurus's uncompromising materialism was seen as interfering at all points with a religious view of the world. Plato had attributed to his Demiurge the role of creating the moving universe "down to the smallest details."[33] The Stoics presented a cosmic, divine *logos* governing providence and human reason alike.[34] Epicurus, in contrast, insisted that the gods had nothing to do with it: the universe was eternal and never had been created; it operated of itself and needed no superintendence.[35]

In the Enlightenment Epicurus's thought was revived and became a major source of inspiration in the development of Western science, directly affecting thinkers as various as Bacon, Gassendi, Descartes, Hobbes, Rousseau, Hume, and Kant. He thus came to be seen as the "inventor of empiric Natural Science" among even idealist philosophers such as Hegel.[36]

Epicurus's denial of any relation of the gods to the material world still generates the ire of intelligent design proponents. Dembski observes that for Epicurus "God or the gods might exist, but they took no interest in the world, played no role in human affairs and indeed could play no role in human affairs, since a material world operating according to mechanistic principles leaves no place for meaningful divine interaction."[37] Phillip E. Johnson argued in *The Wedge of Truth* that materialist views from Epicurus to Gould that allow for the existence of a god or gods as long as they are expelled from the material world can be viewed as an "imperialism . . . founded on materialist premises . . . [that gives] the realm of religion absolutely nothing in the end."[38] Wiker argues in his *Moral Darwinism* that both Epicurus and Gould attempt to make "any deity superfluous" by creating a "two spheres' approach" and removing any divine relation to the material world. It means the elimination of "the Christian cosmos . . . and the Christian moral world as well."[39]

Epicurus, as Marx stressed, rejected the cult of the celestial bodies as gods characteristic of Greek religion and philosophy, especially from the time of Plato, as well as all forms of teleology.[40] Moreover, since humans belonged to nature and were themselves material-sensuous entities, death amounted to dissolution of all material-sensuous connections. Indeed, Epicurus sought to take the fear out of religion (and the afterlife) by denying the existence of the immortal soul, insisting that "death is nothing to us," since there is no longer any sensuous existence and no material reality other than sensuous existence.[41]

Rather than seeing existence as the product of pure chance, as his critics claimed, Epicurus placed the concepts of emergence and contingency at the center of his discussion of the material world, including the changing social world. "Nothing remains for ever what it was. Everything is on the move. Everything is transformed by nature and forced into new paths."[42] Life itself, according to science historian Thomas Hall, translating these views into modern language, was recognized by Epicurus as an emergent consequence of organization; it embodied "action occurring as the *result of organization*," where "the increasingly complex organization of higher life-forms permits the appearance (the emergence) in them of new modes of life, new functions or behaviors, impossible in less organized forms."[43] Thus, the character and behaviors of an organized system, in its totality, cannot be reduced to the operations of its isolated parts.

It is no wonder that today's intelligent design proponents continually evince their dislike for Epicurus. Thus Dembski opens the first chapter of his *No Free Lunch* with complaints about the emphasis that Epicurean philosophy placed on the role of chance, echoing Aquinas and others.[44]

Epicurus and Proto-Evolutionary Theory

Ancient Epicurean materialism was also proto-evolutionary in orientation. It was open to many evolutionary ideas, which were necessary for a materialist perspective, but it lacked a developed theory of the forces that led to evolutionary change. Epicurus taught that life had originally come from the earth. "We are left with the conclusion that the name of mother has rightly been bestowed on the earth, since out of the earth everything is born." Life emerged by spontaneous generation, warmed by the heat of the sun, in early ages when the fertility of the earth was much greater—and not through divine creation.[45] (This early speculative position on the original spontaneous generation of life from non-life was the ancient antecedent to much later scientific theories in the twentieth century, beginning with the Oparin-Haldane hypothesis, which provided a materialist explanation for the emergence of living organisms from the inorganic world.)[46]

Ancient philosophical conceptions related to evolution could be traced to Empedocles (c. 493–433 BCE) and were carried over into the work of Epicurus and his followers. Empedocles presented species as arising from the earth itself. These species included all sorts of monstrous forms (such as "man-faced ox-progeny"), and then through a process of selection those species ill-suited to their environments and incapable of surviving were eliminated. The surviving species were thus restricted to the lasting, normal forms—what in modern scientific parlance could be described as a kind of "normalizing selection."[47]

Although far from being a developed evolutionary perspective in the Darwinian sense, since it failed to encompass the essential argument of evolution by means of natural selection operating on heritable variation across individuals, this approach was nonetheless proto-evolutionary in orientation. In the case of Empedocles it was intermixed with teleological elements. But in its later Epicurean version this proto-evolutionary view was transformed into a strongly materialist argument opposed to the notion of the divine creation of species. Epicurus denied the existence of the more monstrous forms, such as mythological centaurs, depicted by Empedocles, as physically impossible. But Epicurus accepted the idea of wide variation in nature's creations and the survival of the fittest (within a kind of normalizing selection). Those species that were able to continue and perpetuate themselves through the "chain of offspring," forging "the chain of a species in procreation," were those that had special organs that served to protect them from their environment. "But those who were gifted with none of these natural assets . . . were fair game and an easy prey for others, till nature brought their race to extinction."[48]

The human eye, as we have seen, was employed by Socrates as evidence of divine craftsmanship in his original argument from design. The cornea of the eye, as Sedley has noted, was "the only perfectly transparent solid substance known to the ancient Greek world" (as could be seen as early as the fifth century BCE in Empedocles). Reference to the cornea was explicitly incorporated into the argument from design by Cicero in *The Nature of the Gods*, who wrote: "She [Nature] has given our eyes the protection of the most delicate covering membranes, which are at once translucent to give clear sight yet strong enough safely to contain the fluid of the eye."[49] In contrast, Epicurus insisted that it was a mistake to argue that eyes and other organs were purposely designed for use by a creator. "You must not imagine that the bright orbs of our eyes were created purposely, so that we might be able to look before us. . . . In fact, nothing in our bodies was born in order that we might be able to use it, but whatever thing is born creates its own use."[50]

Lacking a developed theory of evolution based on natural selection and yet denying design, Epicurean theory would seem at first glance to rely entirely on blind chance within finite limits, generating absolutely impossible odds against the appearance of such a complex organ as the human eye. Obviously, it would be impossible to imagine that a limited number of purely accidental occurrences in a single world in a restricted span of time would lead to the development of the eye much less the world as we know it. One way this might have transpired is through a kind of Lamarckian evolutionary view, which seems to have been the basis of Epicurus's discussion of human evolution. Thus Epicurus and Lucretius clearly argued with respect to human beings that they had evolved physically and psychologically from more bestial forms. The mechanism behind such descent with modification is unclear, but appears to be linked more to the Lamarckian notion of the inheritance of acquired characteristics in response to the environment; as opposed to Darwinian natural selection or "survival of the fittest."[51]

But Epicurus, according to Sedley, in attempting to provide a materialist explanation of the emergence of the world in all its complexity, also relied on an argument that transformed blind chance into contingency. Thus he adopted assumptions that not only reduced the improbability of the world developing in its present form, but made the appearance of such a world certain. This was what Epicureans called "the power of infinity" associated with the assumptions of (1) infinite space, time, and matter; (2) an infinite number of worlds; (3) a mathematically smallest magnitude (so small as to be partless) that combined in precise ways with other such minimum magnitudes to form atoms (literally uncuttables); (4) a resulting finite number of possible atomic types/shapes derived from the combination of these smallest magnitudes; (5) a largest possible size to a world; and (6) the principle of *isonomia*, or distributive equality between like things. As a result of these mathematical assumptions, together with the basic material postulates of Epicurean philosophy, anything possible was bound to happen in the universe at large, and anything necessary would occur in any given world. Epicurus's swerve, however, went against a strict determinism and guaranteed that no world would be exactly like another. In short, a sophisticated argument of cosmic probability was used to bolster the case for a material explanation of the existing world.[52]

On top of all of this Epicurus provided an account of the development of human society (along with the development of human beings themselves) from an age of stone and wood to that of bronze, and then iron, which also incorporated discussions of the emergence of speech, the advance of mutual assistance, the introduction of fire, and other material changes. Epicurus specifically rejected the Greek mythological view that fire had been given to human beings by Prometheus (a titan and rebel among the gods).[53] Rather, fire was brought down to earth by lightning and kept alive by human beings, or it was discovered to be the result of friction. The historical and sociological content of Epicurus's materialist philosophy thus constituted a rejection of all divine determination of human history. Underscoring this aspect of Epicurus's thought, which emphasized human freedom and fought superstition and state religion, Benjamin Farrington observed: "It is the specific originality of Epicurus that he is the first man known in history to have organized a movement for the liberation of mankind at large from superstition."[54]

Epicurean morality further undermined the agency of the gods by denying foundationalist morality rooted in Platonic ideals, as in the case of justice. In a view that was greatly to influence Marx, he wrote: "If objective circumstances . . . change and the same things which had been just turn out to be no longer useful—then those things were just as long as they were useful for the mutual associations of fellow citizens; but later, when they were not useful, they were no longer just."[55] In other words, with changes in objective conditions, the standards of justice themselves change. Thus, morality was historically shaped and determined by human social practice. Epicurus's morality was at all times rooted in the concept of social contract—a notion he introduced. "The Epicureans," according to Farrington, "were a sort of Society of Friends with a system of Natural Philosophy as its intellectual core."[56]

As Marx stated in *The German Ideology*, "Lucretius praised Epicurus as the hero who was the first to overthrow the gods and trample religion underfoot; for this reason among all church fathers, from Plutarch to Luther, Epicurus has always had the reputation of being the atheist philosopher *par excellence*, and was always called a swine; for which reason, too, Clement of Alexandria says that when Paul takes up arms against philosophy he has in mind Epicurean philosophy alone."[57] Marx himself depicted Epicurus as "the greatest representative of Greek Enlightenment," liberating humans from a teleological world by breaking "the bonds of fate," while providing them with the means to comprehend a universe in transformation. He noted that Epicurus's materialist philosophy carried over into the Enlightenment of the seventeenth and eighteenth centuries, providing it with its humanism and its strength. "Philosophy, as long as a drop of blood shall pulse in its world-subduing and absolutely free heart," Marx wrote, "will never grow tired of answering its adversaries with the cry of Epicurus: 'Not the man who denies the gods worshipped by the multitude, but he who affirms of the gods what the multitude believes about them, is truly impious.' "[58]

4. Enlightenment Materialism and Natural Theology

During the Renaissance numerous long lost works of antiquity were recovered as humanists sought out the missing classics. In 1417 the indefatigable collector of manuscripts Poggio Bracciolini located a copy of Lucretius's *De rerum natura*. This was to become the basis for numerous further copies of Lucretius's poem in the fifteenth and sixteenth centuries. A revival of interest in Epicureanism followed, giving new impetus to materialist thought. Indeed, varying responses to Epicurean materialism came to represent one of the main dividing lines in Enlightenment debates. The very idea of "Enlightenment," as it came to be understood in the eighteenth century in particular, Peter Gay has argued, was to a considerable extent inspired by Lucretius. For "when Lucretius spoke of dispelling night, lifting shadows, or clarifying ideas, he meant the conquest of religion by science. That is precisely how the philosophes used the metaphor." Voltaire had at least six different editions and translations of *De rerum natura* on his shelves.[1]

In the early modern period Epicureanism remained the principal heresy not only for Catholicism but also for emerging Protestantism. Martin Luther in the early sixteenth century claimed that the spread of Epicureanism across Europe was an indication that the end of the world was at hand and accused his opponent Erasmus of belonging to "Epicurus' sty." In 1600, sixteen years before the persecution of Galileo commenced, Giordano Bruno was burned at the stake for spreading heresies, including the Epicurean notion of an infinite universe. Although Bruno's thought contained mystical, hermetic, and pantheistic elements, he argued that matter was the true essence and origin of all things. Bruno's principal contribution to science, according to Thomas Kuhn, had been to demonstrate "the affinity" between the Copernican cosmos and Epicurean atomism, doubly challenging Church doctrine.[2]

Despite growing religious attacks on materialism, the scientific revolution and the emergence of Enlightenment philosophy meant that the old worldviews of God's position in the world were increasingly called into question by rationalist thinkers. In the seventeenth century, Francis Bacon, Thomas Hobbes, and Pierre Gassendi all promoted materialist approaches to science. Bacon, who incorporated Epicurean views into his philosophy, was vehemently opposed to teleology and declared that any argument with respect to nature rooted in final causes was "barren, and like a virgin consecrated to God produces nothing."[3] Hobbes, according to Marx, systematized Bacon, giving greater force to his materialism. Hobbes's friend Gassendi systematized Epicurus's materialism for the new scientific age (attempting at the same time, though less successfully, to bring it into accord with Christianity). John Locke borrowed from Hellenistic epistemology and Gassendi in developing his famous concept of the *tabula rasa*, grounding human reason in experience. René Descartes, while creating a dualistic worldview, systematically excluded God from his physics, where mechanical principles held absolute sway. In the social sciences, figures such as Hobbes, Giambattista Vico, and Jean-Jacques Rousseau were to draw on Epicurus's notion of the social contract and his view of the historical development of human society.[4]

Newton versus Leibniz

Leading British scientists, beginning with Robert Boyle and Isaac Newton, tried to bridge the two worlds of materialist science and Christian religion, incorporating final causes into their arguments, and seeking to make these consistent with the new mechanical philosophy. In an unpublished essay titled "Of the Atomical Philosophy," marked "without fayle to be burn't" upon his death, Boyle indicated his admiration for the atomistic philosophy of Leucippus, Democritus, and Epicurus, which was crucial, particularly as interpreted by Gassendi, for the development of his own corpuscular theory of matter. Nevertheless, the anti-teleological aspects of Epicureanism were to be rejected. Boyle's *Disquisition About the Final Causes of Natural Things* inveighed against the same ancient materialists, arguing that "Epicurus and most of his followers . . . banish the consideration of the ends of things [final causes] because the world being, according to them, made by chance, no ends of anything can be supposed or intended."[5]

In bringing the motions of the planets within a materialist worldview and hence that of science, Newton drew on ancient materialism, and even considered including extracts from Lucretius in his *Principia*. He is recorded to have said: "The philosophy of Epicurus and Lucretius is true and old, but was wrongly interpreted by the ancients as atheism." Nevertheless, Newton relied on the notion of an "intelligent Agent" in his science, in those areas that still seemed to offer no basis for scientific explanation, such as the origins of the solar system.[6]

The question of design and how it was to be understood in the context of the new mechanical philosophy was at the heart of one of the greatest scientific-theological debates of the eighteenth century—between Newton (via Samuel Clarke) and Gottfried Wilhelm Leibniz.[7] In 1715 Leibniz wrote a letter to his friend Caroline, Princess of Wales, questioning Newton's science and philosophy. Leibniz raised various philosophical objections to Newton's work, particularly regarding Newton's claim that God intervened at times to maintain the orbits of the planets. For Leibniz God had created the world as a perfect machine, in which he was simply the "clockmaker," and any necessity of divine intervention to make the world work due to a failure in this mechanism amounted to a heretical claim that "the machine of God's making was imperfect." In Leibniz's philosophy of pre-established harmony God's purpose was perfectly effected in the best of all possible worlds. As he famously put it in his "Preface to the New Essays" (1703–5), "Eyes as piercing as those of God could read the whole sequence of the universe in the smallest of substances. *The Things that are, the things that have been, and the things that will soon be brought in by the future.*"[8] No imperfection therefore was conceivable.

In sharp contrast, Newton in his *Optiks* and elsewhere had given God a number of tasks, including keeping the fixed stars from falling into one another, adjusting the solar system at intervals to correct irregularities in its motions, and preventing the motion of the universe from ebbing.[9] As Leibniz put it in his initial letter to the Princess of Wales, in the view of Newton and his followers "God Almighty wants to wind up his watch from time to time: otherwise it would cease to move. He had not, it seems, sufficient foresight to make it a perpetual motion. Nay, the machine of God's making, is so imperfect, according to these gentlemen; that he is obliged to clean it now and then by an extraordinary concourse, and even to mend it, as a clockmaker mends his work."[10]

Newton's side was taken up by Samuel Clarke, a Newtonian scientist and theologian, who was undoubtedly coached by Newton. The debate took the form of five letters on either side and came to be known as *The Leibniz-Clarke Correspondence*. Clarke challenged the devout Leibniz by claiming that he was actually a materialist in sheep's clothing: "The notion of the world's being a great machine, going on without the interposition of God, as a clock continues to go on without the assistance of a clockmaker; is the notion of materialism and fate, and tends, (under the pretense of making

God a *supra-mundane intelligence*,) to exclude providence and God's government in reality out of the world." Leibniz answered that a world that allowed for "the chance of the Epicureans," to which God was compelled to respond, pointed to a Deity that was "a God only in name." For Leibniz the ideas of the Epicureans were dangerous to piety. In embracing their views of chance and of atoms and void, Newton had gone too far in the direction of crass materialism.[11]

Newton's own position in this debate remains unclear, since there are doubts as to whether Clarke, even with Newton's coaching, adequately represented the latter's views. Newton tended to shy away in his physics from the kind of teleological arguments propounded with metaphysical surety by Leibniz. In contrast to Leibniz's pre-established harmonious design emanating from God, Newton's approach was to explain the physical world as much as possible in materialist-scientific terms. In the case of unknowns, however, Newton allowed God to stand in for an explanation, barring the discovery of a material cause. His theological conceptions, insofar as they entered into his physics, were thus determined by his science rather than the other way around. Even then Newton hesitated to bring God into the picture. Though he certainly believed, as he explained in his "General Scholium" that "this most beautiful system of sun, planets, and comets could only proceed from the counsel and dominion of an intelligent and powerful Being," he avoided whenever possible—where scientific postulates were concerned—any explanation of natural processes as emanating from design. God was the Prime Mover but nature had its own laws. Thus he famously stated with regard to the causes of gravity "I frame no hypotheses," refusing to turn to metaphysical or theological explanations in order to account for gravitational force.[12] "Insofar as Newton was forced," as Harvard geneticist and evolutionary biologist Richard Lewontin has said, "to assume that God intervened from time to time to set things right again," he did so "reluctantly."[13]

This is not to say that Newton in all of his work put science before religion, or that the latter did not influence the former. An equal or greater part of his research over his lifetime was devoted to theology as opposed to physics and mathematics. Newton not infrequently accounted for his scientific efforts in terms of the search for Aristotelian final causes and ultimate confirmation of the existence of a deity. "The main Business of natural Philosophy," he insisted, "is to argue from Phaenomena without feigning Hypotheses, and to deduce Causes from Effects, till we come to the very first Cause [Aristotle's final cause], which certainly is not mechanical."[14] Newton's physics, however, were far removed from the usual argument from design, since his analysis greatly expanded the bounds of science at the expense of the deity, removing the Christian God from realms in which he had previously been seen as dominant, and restricting the divine role to those increasingly remote areas that still had no rational explanation.

Here it is important to recognize that Newton's theological studies were themselves heretical and for that reason remained largely unpublished in his lifetime. Newton strongly embraced Arianism, a fourth-century heresy that had fought with Trinitarians for the soul of Christianity. He adamantly opposed the doctrine of the Trinity. Arianism has similarities to today's Unitarianism in rejecting Christ's full divinity. Newton struggled in his theology to insert a greater rationality into religion. As noted Newtonian scholar Richard Westfall has put it, "The central thrust of his lifelong religious quest was the effort to save Christianity by purging it of irrationalities." In this sense, Newton's approach to religion was undoubtedly affected by the "touch of cold philosophy" associated with the rise of modern science, of which he was to become the leading representative in his day. Newton himself can be seen above all as a rationalist. It was this that led him in his science to remove previous notions of design, while still defending the idea of a divine Creator.[15]

Attempts by today's intelligent design defenders to draw on Newton as a basis for their own arguments, as in the cases of Steve

Fuller, Nancy Pearcey, and Charles Thaxton, invariably come up empty. Although Fuller in his testimony at the Dover Area School District trial (see chapter 1) and in his book *Science vs. Religion?* claimed that "Newton's life appears to imply that the Bible can provide a sure path to great science," he was unable to provide any evidence that Newton's scientific discoveries, such as the theory of gravity, had design arguments as their logical bases—much less that the Bible represented a "sure path" to science.[16]

Indeed, Newton did more than anyone else in his time to obliterate previous notions of design. Pearcey and Thaxton in *The Soul of Science* declare that "in the cosmic order, Newton saw evidence of intelligent design." But they go on to acknowledge that for Newton this merely consisted of arguing backwards by reasoning from effect to cause until arriving at the first or "final" cause, where no natural explanation could yet be found. What Newton achieved in reality through this process, however, was a vast expansion of science at the expense of design—so that the final cause receded almost entirely from the visible world of physics. Ironically, Newton's physics had so thoroughly destroyed traditional Christian theological conceptions of the cause of celestial movements, that even creationist thinkers such as Pearcey and Thaxton observe that Newton as a devout Christian was compelled to try to undo the damage by looking "for avenues to 'fit God in.'"[17] Indeed, it would be more accurate to say that Newton's science, rather than seeking to "fit God in," turned to the God hypothesis in those places the system of nature did not seem to fit, i.e., where gaps appeared in naturalistic explanation. Hence, to the extent that Newton's physics relied on God, it was the "God of the gaps." The "God of the gaps" took over, as C. A. Coulson stated in *Science and Christian Belief* in 1955, "at those strategic places" where science came up short. This type of stopgap argument has provided little real consolation to theists, since "gaps of this sort," he added, "have the unpreventable habit of shrinking" with the advance of science:

Newton, trying to apply his splendid discovery of the law of gravitation
to as many different problems as possible, and finding that although it
would deal with the motion of the moon round the earth, and earth
round the sun, it would not deal with the spinning of the earth round its
polar axis to give us night and day, wrote to the Master of his Cambridge
College, Trinity: "the diurnal rotations of the planets could not be
derived from gravity, but required a divine arm to impress it on them."
This is asking for trouble. For as soon as any one possible scheme is
devised whereby the planets might conceivably have obtained their angu-
lar momentum, the "divine arm" ceases to be needed; science has asserted its ownership over the new territory.[18]

For the Enlightenment as a whole there was no doubt that
Newton's physics had dramatically expanded the realm of science
and materialist explanation at the expense of the Creator. Despite
its intensity, the struggle between Clarke (Newton) and Leibniz had
been about competing visions of a deist compromise between science and religion. A century later, however, materialism had gained
so much ground that deist solutions often appeared quaint by comparison. The motions of the planets, previously seen as governed
exclusively by God's agency, were now understood almost entirely
in mechanical terms. In the eighteenth century the nebular hypothesis of Immanuel Kant and Pierre-Simon Laplace appeared to
remove divine agency even with respect to the origin of the solar
system. Laplace went beyond Newton in demonstrating the
dynamical stability of the solar system, such that God's intervention was no longer required from time to time to set things right.
According to a legendary but largely imaginary story, as related by
Stephen Jay Gould, "Laplace, or so the story goes, gave Napoleon
a copy of his multi-volume *Mécanique céleste* (*Celestial Mechanics*).
Napoleon perused the tomes and asked Laplace how he could
write so much about the workings of the heavens without once
mentioning God, the author of the universe. Laplace replied: 'Sire,
I have no need of that hypothesis.'" Although apocryphal, this

episode was endlessly repeated, and came to stand for the impiety of the new age of science.[19]

Natural Theology: Ray, Paley, Malthus, and Chalmers

Acutely aware that Enlightenment reason had seriously undercut revelation as a basis for knowledge, and unwilling to abandon creationism and thus relinquish the material realm to science, many theists turned to the distinct tradition of natural theology, which sought to find in nature evidence of intelligent design. None other than the great empiricist and deist John Locke wrote in 1695 in *The Reasonableness of Christianity as Delivered in the Scriptures* that, "though the Works of Nature, in every part of them, sufficiently Evidence a Deity; Yet the World made so little use of their Reason, that they saw him not."[20] Arguments for the existence of God from the evidence of nature, mainly in relation to biology, were published in large numbers beginning in the seventeenth century. In Britain John Ray, Samuel Clarke, William Paley, Thomas Robert Malthus, and Thomas Chalmers were among the leading "parson naturalists." In Germany the most important figure in this respect was Hermann Samuel Reimarus.

In his 1691 book, *The Wisdom of God Manifested in the Works of Creation*, Reverend John Ray began with a critique of Epicurus and an attack on the notion of contingency. He viewed what he called "the Atheistik Hypothesis of Epicurus and Democritus" as denying God's wisdom as revealed by creation. His studies of nature were conducted to reveal the marvels of the natural world and how rationally it was organized in accordance with a plan. The design of nature, which would become evident with observation, would make known the providence of God. A vital spirit introduced by God in animals and plants guided their development. This was taken as proof of the active role that God played in nature, as well as an indication of God's wisdom in constructing such a complex, perfect world. Everywhere in nature, Ray affirmed the hand of God at work: the air existed so animals could breathe, and plants grew because God granted them a "Vegetative Soul." Making an analogy to a clock to support his position, Ray stated that a clock shows evidence of a designer, and the organization of nature, more perfect in its design than a clock, indicated that the work of a supreme designer was at hand. The natural theology that Ray presented dominated studies of natural history for nearly two centuries and served as a barrier to the development of evolutionary theory.[21]

A century later, the Archdeacon William Paley, the most influential advocate of natural theology of the late eighteenth and early nineteenth centuries in Britain, extended the argument from design of Ray. In his natural theology, Paley connected the natural and social world. Natural theology was not just an argument about nature; it was an argument regarding the moral universe, which included the economy and the state. In his 1802 book, *Natural Theology—or Evidence of the Existence and Attributes of the Deity Collected from the Appearances of Nature*, Paley argued that proof of God was manifested in the works of his creation. Following the lead of Ray, Paley used a watch analogy—replacing the clock as the high technology of his day—as an argument for design. A watch has a particular ingenuity and its mechanisms work together to tell time as a result of a watchmaker. Thus, he contended, if we could see the contrived design in a watch, the intricate organization and perfection of the operations of nature—such as the marvels of the human eye—should be taken as even more obvious evidence of the work of a grand designer, given how even more wonderful they were than the works of humans. For Paley: "The marks of *design* are too strong to be got over. Design must have had a designer. That designer must have been a person. That person is GOD." Ironically, despite his use of a watch as proof of design, and his pushing of the analogy to the point of referring to a watch that begat other watches, Paley failed to incorporate a sense of time into his conception of nature, which remained essentially static and non-evolutionary in character, excluding emergence. His argument for design focused on what he saw as the irreducible complexity of the natural world, which he thought incapable of materialist explanation. At one point in his *Natural Theology* Paley invokes Adam Smith's "invisible hand" to explain what he took to be evidence of design in nature, but in this case it meant the hand of God.[22]

Paley's more developed views on natural theology and utilitarianism, as expressed in his *Natural Theology*, were foreshadowed in his 1785 book, *Principles of Moral and Political Philosophy*. Here Paley defended existing social hierarchies and property relations. The world was designed in a particular way for beneficial purposes, thus people were not to question who owned the land. The existing system of property rights was to be understood as an "appointment of heaven" for the good of all. God as the "Supreme Proprietor" had consented to the separation of properties only when provision was made for the most elemental needs of the poor.[23] God's plan was just and right. Although this earlier work included an argument to take care of the poor, Paley's *Natural Theology* was to overturn these social concerns. Malthus's influence surfaced in Paley's later book, as he concluded that part of God's design was for every nation to "*breed up* to a certain point of distress."[24]

The same teleological view of "the high purpose of creation" evident in both nature and society was present in Malthus's 1798 *Essay on the Principle of Population*. In this work of political economy and natural theology, Malthus, then a thirty-two-year-old English curate, explained that "we should reason from nature up to nature's God" given that nature was a reflection of the maker's design. He explained that "population should increase faster than food," in accordance with "the gracious designs of Providence, as determined by God." Hardship helped awaken the "Christian virtues" within society. Heads of households who chose to marry without the means to support a family were meant to suffer because they violated "the laws of nature, which are the laws of God," and as a result they were doomed "to starve for disobeying their repeated admonitions." The individual "had no claim of right on society for the smallest portion of food, beyond that which his labour would fairly purchase." Society had no obligation to help those in need, because this would go against the "express commands of God." Malthus ended *A Summary View of the Principle of Population*, his final treatment of the population question, by declaring that the principle of population appeared to accord with "the views of a benevolent Creator" and that the limits this places on human behavior so clearly benefited the well-being of the population that "the ways of God to man with regard to this great law are completely vindicated."[25] The Supreme Being had provided checks—vice and misery—to keep population in a state of equilibrium with the means of subsistence.

For Malthus, the basis of this line of argument was to be found in the discrepancy between the natural geometric rate of growth (e.g., 2, 4, 8, 16) of human population and the natural arithmetic rate of growth (e.g., 2, 3, 4, 5) of subsistence. The natural impediments to food production that resulted in the mere arithmetical increase in food were themselves expressly designed by God. As Malthus put it:

> The necessity of food for the support of life gives rise, probably, to a greater quantity of exertion than any other want, bodily or mental. The Supreme Being has ordained that the earth shall not produce good in great quantities till much preparatory labor and ingenuity has been exercised upon its surface. There is no conceivable connection to our comprehensions, between the seed and the plant or tree that rises from it. The Supreme Creator might, undoubtedly, raise up plants of all kinds, for the use of his creatures, without the assistance of those little bits of matter, which we call seed, or even without the assisting labour and attention of man. The processes of ploughing and clearing the ground, of collecting and sowing the seeds, are not surely for the assistance of God in his creation, but are made previously necessary to the enjoyment of the blessings of life, in order to rouse man into action, and form his mind to reason.

> To furnish the most unremitted excitements of this kind, and to
> urge man to further the gracious designs of Providence by the full culti-
> vation of the earth, it has been ordained that population must increase
> much faster than food.[26]

For Malthus, therefore, the law of the mere arithmetical increase
in human food supply, which constituted the crucial hypothesis in
his theory of population, needed no logical or empirical defense,
since it could be defended as the intelligent design of God.

Yet God in his infinite wisdom had also designed the world so
as to maintain an equilibrium of population and food supply—
through poverty and its attendant misery—and ultimately, if all else
failed, through the dreaded scourge of famine. Malthus's natural
theology thus helped justify class domination and the impoverish-
ment of large sections of the populace.[27] Although employed over
most of his career by the East India College (a training ground for
East India Company officials) Malthus remained a cleric and delivered sermons throughout his life. The recent publication of four of
his sermons has shown that he propounded views based on biblical revelation as well as natural theology. Thus he suggested that
those who sinned against God could expect "terror" in the hereafter (as well as on earth) and that those who used their reason and
attended to God's works in nature could expect to gain insights,
though limited, into his "final causes."[28]

The Scottish divine Reverend Thomas Chalmers was an early
follower and close associate of Malthus. Chalmers was an influential preacher and ecclesiastical reformer within the Established
Church of Scotland and a leader of the schism that resulted in the
creation of the Free Church of Scotland in 1843. This was to
become the most organized group of evangelicals in Britain. Rather
than relying simply on natural theology in its relation to the material world, the Free Church united it with arguments based on revelation and biblical readings, creating what historians have called a
"theology of nature."[29] This dual approach was evident in

Chalmers's own work. He was the author of *On the Power, Wisdom and Goodness of God as Manifested in the Adaptation of External Nature to the Moral and Intellectual Constitution of Man* (1834), the first of the *Bridgewater Treatises* (later to be ridiculed by those in Darwin's circle as the "bilgewater treatises"), a series of eight treatises aimed at combating materialism, funded by a bequest from Francis Henry Egerton, the eighth Earl of Bridgewater, who died in 1829.[30] The *Bridgewater Treatises* constituted the greatest systematic attempt in the nineteenth century to create a natural theology that would dominate all areas of intellectual endeavor. Yet, Chalmers did not confine his activities to natural-theological arguments for design, but also wrote of biblical revelation.

In its New College, the Free Church armed its ministers for combat against materialist and evolutionary theories. As the New College's principal and professor of divinity, Chalmers defended the argument from design against materialists and evolutionary scientists. He fused political economy with natural theology, in an elaborate presentation of how God's hand was evident in the workings of both nature and the economy. For Chalmers "the interposal of a God" and divine miracles were necessary whenever a new genera or species was to come into being.[31]

Chalmers began his *Bridgewater Treatise* by attacking atheists and materialists who tend to

reason exclusively on the laws of matter, and to overlook its dispositions. Could all the beauties and benefits of the astronomical system be referred to the single law of gravitation, it would greatly reduce the argument for a designing cause. . . . If we but say of matter that it is furnished with such powers as make it subservient to many useful results, we keep back the strongest and most unassailable part of the argument for a God. It is greatly more pertinent and convincing to say of matter, that it is distributed into such parts as to ensure a right direction and a beneficial application for its powers. It is not so much in the establishment of certain laws for matter, that we discern the aims or the purposes of intelligence,

as in certain dispositions of matter, that put it in the way of being useful-
ly operated upon by the laws.[32]

For Chalmers both nature and scripture equally led to God.
"Give me the truly inductive spirit to which modern science
stands indebted," he wrote, and it "would infallibly lead. . . to the
firmer establishment of a Bible Christianity in the mind of every
inquirer."[33]

Intelligent design ran deeper than the laws of matter. The world
of trade and the market, Chalmers argued, was "one of the animate
machines of society" and the mark of the "intellect that devised and
gave it birth." The Smithian invisible hand by which self-interest
promoted the general good through the market was, he insisted, the
mark of a "higher agent." The free market was a "natural disposi-
tion"—emanating from God, while the same supreme Deity had
instilled in humanity a strong "possessory feeling." Hence, humanity intervened on behalf of the poor, as in the Poor Laws, in vain
arrogance, defying the will of God. "Capital ever suits itself, in the
way that is best possible, to the circumstances of the country—so as
to leave uncalled for, any economic regulation by the wisdom of
man; and that precisely because of a previous moral and mental
regulation by the wisdom of God." Indeed, if there was any proof
of "the hand of a righteous Deity" it was to be found in the "mechanism of trade."[34]

Hume's Critique of Natural Theology

In the ceaseless battle between materialism and creationism in the
Enlightenment period it was the skeptical view of David Hume that
most powerfully challenged the argument from design. Hume's critique was most fully manifested in his treatise, *Dialogues
Concerning Natural Religion*, published posthumously in 1779.
But an earlier, fundamental critique of intelligent design was to be
found in Hume's *Enquiry Concerning Human Understanding*
(1748). Here he provided as the core of a dialogue on natural reli-

gion an imaginary speech by Epicurus, defending himself before
the Athenian population against charges of impiety. Hume's
Epicurus began by noting that his accusers found "marks of intelligence and design" in the "order of nature," so they thought it
"extravagant to assign for its cause, either chance, or the blind and
unguided force of matter"—as he himself had asserted. Epicurus's
response to this accusation was that the "religious hypothesis," or
argument from design, requires reasoning from a particular effect
in the world to an antecedent cause. Yet no specific qualities could
be reasonably inferred from given effects and attributed to any particular cause other than those that were requisite to produce the
given effects. Hence, no reasoning backwards from effect to cause
in the natural world could arrive with certainty at *intelligence* and
design, much less the notion of supreme beings as a cause. As
Hume's Epicurus concluded his defense:

> While we argue from the course of nature, and infer a particular intelli-
> gent cause, which first bestowed, and still preserves order in the universe,
> we embrace a principle, which is both uncertain and useless. It is uncer-
> tain; because the subject lies entirely beyond the reach of human experi-
> ence. It is useless; because our knowledge of this cause being derived
> entirely from the course of nature, we can never, according to the rules of
> just reasoning, return back from the cause with any new inference, or
> making additions to the common and experienced course of nature,
> establish any new principles of conduct and behaviour.[35]

Hume's work, *Dialogues Concerning Natural Religion*, relied
on the foregoing argument but extended it to the question of the
role of analogy. The argument from design, Hume suggests via his
character Philo in these *Dialogues*, relied on nothing more than an
"analogy of nature," proceeding in wide "steps" from one analogy
to another. The structure that such an argument invariably took
was that of inferring by analogy some instance of design in nature,
and then attributing that particular design—on the assumption that
all designs must have a designer—to a Supreme Designer or

Intelligence from which it arose. For example, "the world . . . resembles a machine, therefore it is a machine, therefore it arose from design," consequently there must be a Supreme Designer. But other equally plausible analogies, Hume claimed, could be drawn to explain the generation of the world and the existence of order without recourse to God. Thus, referring explicitly to Epicurean arguments, Philo in Hume's *Dialogues Concerning Natural Religion* pointed to the possibility of the generation of the world from seeds, matter in motion, the emergence of order from the blind interactions of material forces, and "the powers of [the] infinite." Out of this could arise organization:

> A tree bestows order and organization on that tree which springs from it, without knowing the order: an animal, in the same manner, on its off-spring: a bird, on its nest: And instances of this kind are even more frequent in the world, than those of order, which arise from reason and contrivance. To say that all this order in animals and vegetables proceeds ultimately from design is begging the question; nor can that great point be ascertained otherwise than by proving *a priori*, both that order is, from its nature, inseparably attached to thought, and that it can never, of itself, or from original unknown principles, belong to matter.[36]

Philo, standing for Hume's own views, ended with a skeptical position, endorsing neither materialism nor religion. Yet, there can be no doubt toward which side Hume himself ultimately leaned. On his deathbed, Hume, as recounted by his friends James Boswell and Adam Smith, steadfastly refused to embrace religion, taking comfort instead in Epicurus's materialist views (via Lucretius and Lucian).[37]

Dr. Pangloss and Natural Religion

Nowhere was intelligent design held up to more ridicule in the eighteenth century than in Voltaire's *Candide*. There Voltaire's character Dr. Pangloss, for whom Leibniz was the chief model, is introduced as follows:

Pangloss taught metaphysico-theologo-cosmolo-nigology. He proved incontestably that there is no effect without a cause, and that in this best of all possible worlds, his lordship's country seat was the most beautiful of mansions and her ladyship the best of all possible ladyships.

"It is proved," he used to say, "that things cannot be other than as they are, for since everything was made for a purpose, it follows that everything is made for the best purpose. Observe: our noses were made to carry spectacles, so we have spectacles."[38]

5. Marx's Critique of Heaven and Critique of Earth

The Critique of Heaven

"Christianity," Karl Marx observed, "cannot be reconciled with reason [as embodied in Enlightenment science] because 'secular' and 'spiritual' reason contradict each other."[1] Marx was a strong critic of teleology and the argument from design, which he saw as alienated attempts to provide a rational basis in nature for God's dominion on earth, thereby justifying all earthly dominions. He sided with the materialist critique of intelligent design emanating from Epicurus, whom he called in his doctoral dissertation "the greatest representative of Greek Enlightenment."[2] Marx therefore stands next to Darwin and Freud as a target for today's intelligent design proponents—who trace the intellectual sins of all three ultimately to Epicurus.[3]

For Marx the critique of religion was the indispensable starting point for a broader critique of an "inverted world" for which religion was both the "general theory" and the "encyclopedic compendium." As he stated in 1844 in his "Introduction to a Critique of Hegel's Philosophy of Right": "The criticism of heaven turns into the criticism of earth, the *criticism of religion* into the *criticism of law* and the *criticism of theology* into the *criticism of politics*."[4] It was the critique of religion that made philosophy and science (and with this the critique of political economy) possible. This also described the progression of Marx's own thinking.

Marx came from a mixed Jewish-Lutheran-Deist heritage. Both of his maternal and paternal grandfathers were rabbis, and almost all of the rabbis of Trier from the sixteenth century on were his ancestors. But his father, Heinrich Marx, converted to Lutheranism by 1817, the year before Marx's birth, so that he could continue his profession as a lawyer in the Prussian state, which would otherwise have barred him from employment. Heinrich Marx was to become a devoted deist, described by Edgar von Westphalen (Karl Marx's future brother-in-law) as a "Protestant *à la* Lessing." He embraced the Enlightenment, could recite Voltaire and Rousseau by heart, and urged his son to "pray to the Almighty" and "to follow the faith of Newton, Locke and Leibniz." Not as much is known about the beliefs of Marx's mother, Henrietta. She seems to have been more attached to her Jewish beliefs, partly in deference to her parents' feelings, and was not baptized until 1825 (a year after Karl) upon the death of her father. The young Marx also came under the tutelage of the Baron Ludgwig von Westphalen (his future father-in-law) who introduced him early on to the ideas of the utopian socialist Saint-Simon.

Marx was educated at the Friedrich Wilhelm Gymnasium (High School) in Trier, a former Jesuit school in which four-fifths of the students were Catholic. In 1835 at the age of seventeen he was required to write three essays for his school-exit examination. One had to be devoted to a religious subject and Marx wrote on "The Union of Believers with Christ, According to John 15: 1–14, Showing its Basis and Essence, its Absolute Necessity, and its Effects." The paper presented the Lutheran Trinitarian argument on the necessity of the union with Christ as the goal of history. Marx concluded his paper by stating that "union with Christ bestows a joy which the Epicurean strives vainly to derive from his frivolous philosophy or the deeper thinker from the most hidden depths of knowledge." This early focus on Christ versus the

Epicureans and other philosphers suggests that even as an adolescent Marx was already interested in Epicurus's materialism and its critique of design, pointing to his doctoral dissertation six years later on Epicurus in which he was to reverse the position of his early school paper and embrace the critique of design. Marx's school essay on religion was written in the same year as David Strauss published his *Life of Jesus*, which was to constitute the starting point of the Young Hegelian critique of religion (and the same year as the introduction of the railway into Germany).[5]

Following his early school papers, the next major extant record emanating from Marx's pen is his remarkable letter to his father, written from Berlin in November 1837. Here we find Marx struggling over the "grotesque craggy melody" of Hegel's philosophy, which he absorbed completely but also resisted in part due to its idealistic content. "If previously the gods had dwelt above the earth," he wrote, "now [in Hegel] they became its centre." Here was a philosophy "seeking the idea in reality itself." But despite its obvious power over his thought, Marx felt that he had been delivered "into the arms of the enemy" and that he "had made an idol of a view" he "hated." At the same time he joined the Young Hegelian "Doctors' Club," which endlessly discussed Hegel's philosophy and the critique of religion.[6]

In the very midst of his struggles over Hegelian philosophy Marx turned to "positive studies," investigating the works of both Francis Bacon and the German natural theologian Hermann Samuel Reimarus. The long-term impact of Bacon on Marx's thinking cannot be doubted. Marx saw Bacon as the modern materialist counterpart of the ancient atomists Democritus and Epicurus.[7] Marx and Darwin within a few years of each other in the late 1830s and early 1840s both explicitly adopted Bacon's anti-teleological view, drawn from the ancient materialists, that any concept of nature rooted in final causes was "barren, and like a virgin consecrated to God produces nothing."[8] Marx was undoubtedly strongly influenced by Hegel's extensive treatment, in his *History of Philosophy*, of Bacon's critique of final causes (for example, the notion that the bee is "provided with" a stinger for protection) as opposed to efficient causes. In presenting Bacon's critique of intelligent design, Hegel depicted him as the modern representative of an argument that "has the very merit [of opposing "superstition generally"] which we met with in the Epicurean philosophy."[9] In this way, the great millennial struggle between materialism and idealism, between science and teleology, with regard to the interpretation of nature, impressed itself early in Marx's thought via Bacon, and was reinforced by his studies of Hegel. The Enlightenment materialism of the eighteenth century, as Engels put it, "posited Nature instead of the Christian God as the Absolute confronting man." Such materialism derived from the rejection within science of both the argument from design of Christian religion and all idealistic theories that relied on teleological arguments. As Engels cogently expressed it (quoted earlier in the introduction to this book):

> "Did god create the world or has the world been in existence eternally?" The answers which the philosophers gave to this question split them into two great camps. Those who asserted the primacy of spirit to nature and, therefore, in the last instance, assumed world creation in some form or other—(and among the philosophers, Hegel, for example, this creation often becomes still more intricate and impossible than in Christianity)— comprised the camp of idealism. The others, who regarded nature as primary, belong to the various schools of materialism. These two expressions, idealism and materialism, primarily signify nothing more than this; and here also they are not used in any other sense.[10]

If the issue of materialism versus idealism necessarily arose in Marx's 1837 studies of Bacon, this was no less true of his 1837 readings from Reimarus. Hermann Samuel Reimarus was best known in Marx's day for his posthumously published *Wolfenbüttel Fragments* of 1774–78, drawn from his *Apology, or Defense for the Reasonable Worshippers of God.* Representing a rationalist, deist criticism of the accuracy of biblical revelation with regard to Christ and a denial of his divinity (Reimarus called Christ a "secular savior"), the *Fragments* created a furor in Germany—not unlike the reception of David Strauss's *Life of Jesus* in the following century.[11] In his lifetime, however, Reimarus was known principally for his work on logic and—more significantly for Marx—for his two major works on natural theology and the instincts of animals: his 1754 *The Principal Truths of Natural Religion Defended and Illustrated, in Nine Dissertations: Wherein the Objections of Lucretius, Buffon, Maupertuis, Rousseau, La Mettrie, and Other Ancient and Modern Followers of Epicurus Are Considered, and Their Doctrines Refuted* and his 1760 *Triebe der Thiere,* or *Drives of Animals.*[12]

Reimarus was a follower of the English natural theologian John Ray and had written a brief treatise as early as 1725 promoting Ray's argument from design. The influence of Ray is evident throughout Reimarus's *Principal Truths of Natural Religion.* He transformed Ray's clock metaphor into a watch metaphor nearly half a century before William Paley more famously employed the watch metaphor in his *Natural Theology* (1802). As Reimarus wrote:

> Suppose a Hottentot who knows nothing of the use of a watch, was shewn the inside, the spring, chain, wheels, in short, all its parts and the disposition of them; nay let him be instructed by a watch-maker, so that, in time, he may be able to make a watch; yet I affirm, that the Hottentot, if he is not made acquainted with the use of a watch, does not know what a watch is. He knows it not essentially; he is ignorant of its design and entire construction. For if the use of it had not been previously conceived in the mind of the artist who made a watch, as something sensible, such a machine would never have been made, nor have been disposed and constructed in such a manner.

Reimarus used this argument to infer that just as a watch was a machine designed by humanity for its own use, so the entire machinery of the inanimate world must have been designed by God and for a purpose: for use by animate beings.[13]

The principal thrust of Reimarus's *Principal Truths of Natural Religion* was to counter the ancient Epicurean critique of intelligent design and its modern representatives. Thus he argued against Epicurean "blind chance" and in favor of God's "wisdom and design." The ultimate crime of Epicurus's philosophy, according to Reimarus, was to "banish God into the *Intermundia*," leaving him with no relation to the world. In the first five of the nine "dissertations" that made up this work Reimarus principally concerned himself with attacking Epicurus's own arguments, while in the remaining four dissertations he addressed the modern followers of Epicurus (such as Buffon, Maupertuis, Rousseau, and La Mettrie). Arguing against the notion of the spontaneous creation of life from the earth, he declared in direct opposition to Epicurus's view: "The origin of men and other animals from the earth cannot be accounted for in a natural way. . . . [T]he earth has no title to be called the general mother of us all."[14]

It was in the fifth dissertation of his *Principal Truths of Natural Religion* that Reimarus most effectively advanced what he called the "general proof" of final causes, focusing on the innate drives of animals, and distinguishing these from human knowledge derived from experience. Animals, he argued, obtained the rationality evident in these innate drives directly from God rather than material causes. Writing, for example, of bees he stated: "Certainly no part of Nature shews greater appearances of a Superior Direction than the Bees, which not only form their sexangular cells in the most regular and just dimensions, but go about it as if they were well versed in the sublimest parts of geometry and fluxions." In contrast, "When men first come into the world, they have very few or no ideas, and have no skill or ability to put any plan in execution, but acquire them by invention and exercise...attain[ed] only by repeated trials and long practice." Indeed, human beings have "for . . . many thousand years, been labouring with united strength in the invention of their arts, which have been but slowly brought to the present degree of perfection; and yet we cannot be said to execute what is necessary for our station in so perfect a manner as every animal, in its way, does immediately after its birth." For Reimarus this was sufficient to establish the truth that animals "owe all their skill to a superior Intelligence." Playing on the ancient materialist proposition, advanced most consistently by Epicurus, that "nothing comes from nothing," Reimarus argued that "from nothing, nothing can be conceived or invented"—hence the innate drives of animals had to be attributed to "the over-ruling Wisdom of their Creator."[15]

Six years later in his *Drives of Animals* Reimarus expanded this argument of the "fifth dissertation" of his *Principal Truths of Natural Religion* into a more general animal psychology. Here the argument from design is pushed further into the background and a more scientifically modeled argument is constructed, though Reimarus never abandoned his natural-theological views. "For the mature Reimarus," Julian Jaynes and William Woodward wrote in the *Journal of the History of the Behavioral Sciences*, "the explanation of animal behavior is not" to be found in "incorporeal knowledges implanted either by God or experience, but . . . innate physiological organizations called drives." Consequently, he has been called "the originator of the concept of drives" in psychology.[16]

Reimarus's theory of drives was largely ignored by psychology until the twentieth century but had an important impact on Marx, who frequently employed the psychological component of Reimarus's theory of drives in his own distinctions between human beings and animals. Inspired in part by Reimarus, Marx used the comparison of bees as natural architects to bring out the distinctiveness of human labor. "A spider conducts operations which resemble those of a weaver, and a bee would put many a human architect to shame by the construction of its honeycomb cells. But what distinguishes the worst architect from the best of bees is that the architect builds the cell in his mind before he constructs it in wax." It is clear that Marx, as he himself indicated, studied Reimarus closely, including his natural theology and his critique of

Epicurean materialism: issues that were central to Marx's analysis from the beginning.[17] Marx, however, would have had little patience with Reimarus's natural theology. Thus he was to refer with disdain to "the earlier teleologists" for whom "plants exist to be eaten by animals, and animals exist to be eaten by men."[18]

Marx preferred Newtonian deism both to the natural theology of Reimarus's *Principal Truths* and the "best of all possible worlds philosophy" of Leibniz. With regard to the famous seventeenth-century debate between Samuel Clarke (representing Newton) and Leibniz, Marx clearly sided with Clarke-Newton's greater adherence to scientific principle—writing "Bravo, old Newton!" in response to Newton's position in his *Principia* (quoted in the *Leibniz-Clarke Correspondence*), in which he forcefully denied that God was "the soul of the world" as opposed to having dominion over souls as the "Universal Ruler." Newton's position was a partial recognition (in the natural realm) of the separation of the magisteria of science and religion.[19]

These concerns regarding materialism and design carry over into Marx's doctoral dissertation. His dissertation, *The Difference Between the Democritean and Epicurean Philosophy of Nature*, was completed and accepted in 1841. However, he began his work on it in 1839, when he commenced his seven notebooks on Epicurean philosophy. His dissertation also included an appendix, "Critique of Plutarch's Polemic Against the Theology of Epicurus," of which we have only a couple of fragments plus the notes to the appendix. (The last two chapters of the first part of the dissertation are also missing from the extant document, except for a part of the notes for the missing chapter 4.)

Marx's dissertation, despite its title, was concerned relatively little with the philosophy of Democritus, which was mainly a springboard for his analysis of Epicurus. As philosopher Paul Schafer has explained, "The dissertation's substantive core, that is, its atomist or materialist content, is Epicurean, while its analytical approach, that is, the dialectical method utilized to think those core ideas through, is Hegelian. The result is a fascinating hybrid that provides an illuminating picture of the genesis of Marx's philosophical view": the struggle between materialism and idealism that was to govern his thought. Marx strongly admired Epicurus's materialism, his "dialectical atomism" (as Schafer puts it), his critique of teleology and determinism, and above all his philosophy of freedom. Perhaps nothing so drew Marx to Epicurus as much as the latter's statements (strung together by Marx from ancient sources): "'It is chance, which must be accepted, *not God*, as the multitude believe.'... 'It is a misfortune to live in necessity, but to live in necessity is not a necessity. On all sides many short and easy paths to freedom are open.... It is permitted to subdue necessity itself.'"[20]

As Engels later wrote: "While classic Greek philosophy in its last forms—particularly in the Epicurean school—led to atheistic materialism, Greek vulgar philosophy led to the doctrine of a one and only God and of the immortality of the human soul."[21] In the debates regarding natural science versus religion in his lifetime, Marx identified with the struggles and dilemmas that Epicurus confronted and the materialist, empiricist tradition to which he gave rise. Hence, Arend Th. van Leeuwen, a theologian, pointed out in relation to Marx's dissertation, "In a sense, Epicurus acts as Marx's double. Every time the name Epicurus is mentioned, we are to think of Marx reflecting his own problems in the mirror of Greek philosophy."[22]

At the core of materialism was a critique of the notion that the rationality of the world was to be attributed to the gods. Hence, Marx's doctoral dissertation on Epicurus was both a treatment of materialist dialectics and a critique of religion. In his "Foreword" to what was intended to be a published version of his thesis, Marx identified Epicurus with Prometheus (both bringers of light) and contended: "The confession of Prometheus: 'In simple words, I hate the pack of gods' [Aeschylus, *Prometheus Bound*], is its [philosophy's] own confession, its own aphorism against all heavenly and earthly gods who do not acknowledge human self-consciousness as the highest divinity. It will have none other beside."

Justifying the inclusion of the appendix to his dissertation, Marx wrote: "If a critique of Plutarch's polemic against Epicurus' theology has been added as an appendix, this is because this polemic is by no means isolated, but rather representative of a species, in that it most strikingly presents in itself the relation of the theologising intellect to philosophy." By attempting to promote the religious morality and the argument from design and to polemicize against Epicurus on those bases Plutarch had brought "philosophy before the forum of religion." Marx went on to side explicitly with Hume in declaring that philosophy with its rational approach to nature, and not the "theologising intellect" of natural theology, is the rightful king of the realm of reason.[23]

Plutarch—who lived into the second century—was the senior of two priests of Apollo at the Oracle of Delphi and "a representative of religious Platonism during the early part of the Christian era."[24] He was a strong critic of Epicurus, on the grounds that the latter had removed the necessary fear of God. It was terror of the afterlife that above all bound humanity to God. As Marx put it, Plutarch was a spokesperson for the doctrine that "justifies the terrors of the underworld for the sensuous consciousness. . . . In fear, and specifically in an inner fear that cannot be extinguished, man is determined as an animal." At the same time Plutarch advanced the argument of benign providence (even in the most terrible acts) as proof of God's existence. For Plutarch, Epicurus was to be castigated for transforming the gods into distant beings comparable to the "Hyrcanian [Caspian Sea] fish" from which no harm or advantage could be obtained.[25]

Marx's critique of Plutarch both in the main text of his dissertation and in its appendix is thus of great importance in understanding his critique of religion, and in his response to the argument of design specifically. Marx had nothing but contempt for Plutarch, who in addressing in his biography of Marius the battle between the Romans and the German Cimbri tribes in 101 BCE near Vercelli provided what Marx called "an appalling historical example" of how a religious morality rooted in the fear of all-powerful deities violated all conceivable humanity:

> After describing the terrible downfall of the Cimbri, he relates that the number of corpses was so great that the Massilians [i.e., citizens of the Greek colony and city-state Massilia, now Marseilles] were able to manure their orchards with them. Then it rained and that year was the best for wine and fruit. Now, what kind of reflections occur to our noble historian in connection with the tragical ruin of those people? Plutarch considers it a moral act of God, that he allowed a whole, great, noble people to perish and rot away in order to provide the philistines of Massilia with a bumper fruit harvest. Thus even the transformation of a people into a heap of manure offers a desirable occasion for a happy reveling in [religious] morality![26]

Hence, for Plutarch a bumper crop of wine and fruit resulting from the rotting bodies of the vanquished Cimbri was itself an argument for the rationality of nature arising from divine providence. Plutarch's God was for Marx a "degraded God," and Plutarch himself a spokesperson for "the hell of the populace."

In refutation of Plutarch, Marx, in his appendix, rejected "proofs of the existence of God," since these were in reality their opposite: "proofs of the existence of essential human self-consciousness." Indeed, "the country of reason," he declared, "is for God in general, a region in which he ceases to exist"—since this is the exclusive realm of humanity. Contra to Plutarch, Marx quotes from the French materialist and Epicurean Baron d'Holbach's *System of Nature*: "Nothing ... could be more dangerous than to persuade man that a being superior to nature exists, a being before whom reason must be silent and to whom man must sacrifice all to receive happiness."

Both the later Friedrich Wilhelm Joseph von Schelling and Hegel come under attack in Marx's appendix for their theological views. Schelling is seen as abandoning his earlier conception of human freedom in concluding in his later work that God "is the real foundation of our cognition." Hegel is condemned for turning all previous theological demonstrations "upside down" in order to try to demonstrate God's existence in the opposite fashion from traditional Christian theology. Previously, natural accidents and miracles were considered the proofs of God's existence. Now Hegel, in line with natural theology, purported to demonstrate the same thing with the reverse argument. Simply because "the accidental does *not* exist, God or the Absolute exists." In other words, the proofs of God were to be found not in natural accidents or in miracles but in evidence of divine necessity.

Responding to such alleged proofs of God's existence, including the argument from design, Marx pithily declared: "Lack of reason is the existence of God." Conversely, the historical development of self-consciousness in the material world is the reasoned existence of humanity. "It is precisely Epicurus who makes the form of consciousness in its directness, the being-for-self, the form of nature. Only when nature is acknowledged as absolutely free from conscious reason [i.e., from the externally imposed rationality of a deity] and is considered reason in itself, does it become entirely the property of reason," or the self-conscious world of humanity.[27]

In this Marx broke sharply with Hegel, who had proclaimed in holy terms that his *Logic* was nothing but "the exposition of God as He is in His eternal essence before the creation of the world and man."[28] At the heart of Hegel's entire philosophy, Marx and Engels were to state in *The Holy Family*, was "the *speculative* expression of the Christian-Germanic dogma of the antithesis between *Spirit* and *Matter*, *God* and the *world*." In his *Critique of Hegel's Philosophy of Right* Marx went so far as to label Hegel's *Logic* a "*Santa Casa*" or "Holy House"—the name with which, as van Leeuwen pointed out, the Roman Catholic Inquisition in Madrid "sanctified its prison" and chamber of terror.[29]

It has been customary to see Marx's critique of religion and of Hegelian philosophical idealism as only developing as a result of his encounter with Ludwig Feuerbach's prior critique of the Hegelian system. However, Marx's critique of the "theologising intellect," which was to find its most powerful expression in his introduction to *Critique of Hegel's Philosophy of Right* in 1844, was essentially complete by the time he submitted his doctoral dissertation in early 1841—the very year that Feuerbach's *Essence of Christianity* was published.[30] Moreover, Feuerbach's *Preliminary Theses on the Reform of Philosophy*, which was to have a more direct impact on Marx's thinking, did not appear until 1842. It would be more correct to argue, therefore, that Marx's critique of religion developed independently of and alongside Feuerbach's critique, which added force to Marx's views.[31]

Nonetheless Feuerbach's naturalistic rejection of Hegel's idealist philosophy exerted a powerful influence on Marx. For Feuerbach, speculative philosophy in its most developed form, the Hegelian system, represented the alienation of the world of sensuous existence to which human reason was materialistically bound. It replicated, in the name of philosophy rather than theology, the religious estrangement of human beings from nature. Hegel had presented the world as developing in inverted form "from the ideal to the real." In contrast, "all science," Feuerbach insisted, "must be grounded in *nature*. A doctrine remains a *hypothesis* as long as it has not found its *natural basis*. This is true particularly of the *doctrine of freedom*. Only the new [materialist] philosophy will succeed in *naturalizing* freedom which was hitherto an *anti-hypothesis*, a *supernatural hypothesis*."[32]

Marx's *Critique of Hegel's Philosophy of Right*, which was published in 1844 in Paris in the *Deutsch-Französische Jahrbücher* (*German-French Annals*), has been called "the Magna Carta of the Marxist critique of religion."[33] It is here that Marx declared:

> *Man makes religion*, religion does not make man. Religion is indeed the self-consciousness and self-esteem of man who has either not yet won through to himself or has already lost himself again. But *man* is no abstract being squatting outside the world. Man is *the world of man*, state, society. This state and this society produce religion, which is an

inverted consciousness of the world, because they are an *inverted world*. . . .
[Religion] is the *fantastic realization* of the human essence since the
human essence has not acquired any true reality. The struggle against
religion is therefore indirectly the struggle against *that world* whose spiritual *aroma* is religion.

Religious suffering is at one and the same time the *expression* of real
suffering and a protest against real suffering. Religion is the sigh of the
oppressed creature, the heart of a heartless world and the soul of soulless
conditions. It is the *opium* of the people.

Marx here demonstrates a real sympathy for religion "as the
expression of real suffering" and as a necessary solace for the
oppressed. The latter do not have the same access to other means
of consolation, such as opium, available to the wealthy, and have
not yet learned to revolt against the inverted world of which religion is the *fantastic* manifestation. Reversing the position he
adopted as an adolescent for his school paper on the "Union of the
Faithful with Christ," Marx argued that "the abolition of religion as
the *illusory* happiness of the people is the demand for their *real*
[material] happiness." "Thus the criticism of heaven turns into the
criticism of earth."[34]

The Critique of Earth

Marx's critique of religion was geared at all times to the needs of a
humanist, materialist, and scientific understanding of the world.
The critique of religious alienation led to the critique of human-
worldly alienation by means of two dialectical movements: (1) a critique derived from Epicurus and Feuerbach of religion as the alienation of the human world, and thus an inversion of human freedom—a critique that also extended from theology to idealist philosophy (as in the case of Hegel); and (2) a critique of purely contemplative materialism/humanism as empty abstractions, insofar as
they were not simply presuppositions for a critique of earth (i.e.,
material-historical reality).

Hence, atheism itself, so long as it remained in the ether of Feuerbach's contemplative realm, was insufficient and devoid of genuine meaning, other than as a first step in the development of a humanist philosophy. Atheism as an ideal, Marx insisted, was "for the most part an abstraction." It was "a *negation of God*, through which negation it asserts the *existence of man*." It thus constituted mere "theoretical" humanism.[35]

As a materialist, Marx opted *not* to invest in the abstraction of God and religion. At the same time he did not attempt to disprove the supernatural existence of God, since that transcended the real, empirical world and could not be answered, or even addressed, through reason, observation, and scientific inquiry. Instead, he forged a practical atheism through his scientific commitment to a historical materialist approach for understanding reality in all of its dimensions. The practical negation of God and the affirmation of humanity and science demanded an active movement for revolutionary social change, the real appropriation of the world to pursue human development—the growth and expansion of human capabilities—and freedom.

Marx's critique of religion was thus never about the supernatural existence (even in negation) of God, but about the affirmation of the material world, the world of human beings, of reason and science—all of which required the displacement of "religion" as "the devious acknowledgement of man, through an intermediary."[36] Thomas Dean was therefore correct when he wrote in his *Post-Theistic Thinking*:

> Agreeing with the Aristotelian and Hegelian observation that contraries belong to the same genus, Marx views atheism as nothing more than an ideological contrary to religion. Hence it does not lead to a radical break with a religious way of thinking. Atheism looks more like a "last stage of theism, a negative recognition of God" than the theoretical foundation for a positive, this-worldly philosophy of man. It gives rise inevitably to the desire to supplant the God thus denied by a correspondingly elevated or

deified concept of man. . . . It is only by a second act of transcendence, by transcending the mediation of humanism via atheism, "which is, however, a necessary presupposition," that the possibility opens up of a "positive humanism, humanism emerging positively from itself." The basis of Marx's atheism and of his secular metaphysics is not therefore a set of philosophical arguments or speculative disproofs of the existence of God. That would be an ideological foundation as theological in character as theology itself. It is, rather, an independently formulated humanism that stands in immediate or unmediated fashion on its own feet.[37]

Marx's dialectical position that viewed religion as the source of "an illusory happiness," made necessary by the impossibility of "real happiness," meant that it was possible to recognize the alienated humanity in religion itself. Thus he was capable of not only referring to religion as "the heart of a heartless world," but also of making such statements as: "After all we can forgive Christianity much, because it taught us the worship of the child."[38] Compared to this, as Marx observed in his *Theses on Feuerbach*, a crude atheism that sought to establish itself alongside traditional religion "as an independent realm in the clouds" had relatively little to offer. The critique of religion was therefore socially meaningful only to the extent that it went beyond abstract atheism and contemplative materialism and gave rise to an atheism on the ground rooted in "revolutionary practice."[39]

Marx's early critique of religion and of speculative philosophy was to form the basis of his later critique of ideology, specifically the ideology of bourgeois society. Ideology thus became a more general case of the same inversion of ideas and the material world that characterized the alienated condition of religion: "The ruling ideas," Marx and Engels wrote in the *German Ideology*, echoing Marx's earlier critique of the "theologising intellect," "are nothing more than the ideal expression of the dominant material relations, the dominant material relations grasped as ideas."[40]

Marx often referred to the Protestant Reformation, and specifically Lutheranism in the German context, as representing the new religious garment that clothed the rising bourgeois society. Thus he ironically pointed to Martin Luther's argument on the existence of a universal world of plunder as evidence of God's design. As Luther put it, "God uses knights and robbers as his devils to punish the injustice of merchants." In this way, according to Luther, "unchristian thieving and robbing" on all sides could be seen as pointing to the eventual coming to be of "God's final word." Thus for Luther—as Marx clearly meant his readers to understand—God's rationality was displayed even in what Hobbes had called "the war of all against all" of bourgeois society.

In Marx's *Capital*, money, commodities, and capital were all seen as taking on the form of God in bourgeois society, and profit, rent, and interest formed a new "Trinity." Marx compared the "fetishism that attaches itself to the products of labour" to the "misty realm of religion" where "the products of the human brain appear as autonomous figures endowed with a life of their own."[41] The parallels between the critique of religion and the critique of capital in Marx's thought are thus endless.

Yet Marx also continued to confront religion (including the argument from design) more directly due to its intrusions in the realms of morality and science. Morality was to be judged not in either foundationalist or relativist terms, but in terms of radical historicism, where moral conditions evolve with the material needs of human communities—a view that could be traced to Epicurus. There was no ultimate, divine moral order for society. Marx therefore attacked all notions of "*mystical tendency*, the *providential aim* ... providence." He rejected all foundationalist morality emanating from religious final causes, insisting instead that human beings were "the actors and authors of their own drama."[42]

Denouncing narrow religious morality and its effects on the development of political economy, Marx observed in *Capital* that "most of the population theorists are Protestant clerics ... Parson Wallace, Parson Townsend, Parson Malthus and his pupil, the arch-Parson Thomas Chalmers, to say nothing of lesser reverend scribblers in this line. . . . With the entry of 'the principle of population' [into political economy], the hour of the Protestant parsons struck."[43] The main objection to such thinkers was that they had departed from the principles of science by allowing the arguments of natural theology and religious morality to intrude into the science of political economy, as part of a defense of the ruling-class order. "The Malthusian theory," the young Engels wrote in 1844, was "the economic expression of the religious dogma of the contradiction of spirit and nature and the resulting corruption of both."[44] In his 1786 *Dissertation on the Poor Laws*, Reverend Joseph Townsend, as Marx noted in the *Grundrisse*, supplemented fear as a motive for Christian religion with hunger as a motive for bourgeois industry (both constituting evidence of natural law and God's design). "Hunger," Townsend wrote, "is not only a peaceable, silent, unremitted pressure, but, as the most natural motive to industry and labour, it calls forth the most powerful exertions."[45]

For Marx, Malthus, like Townsend before him, was guilty of "clerical fanaticism."[46] Although Malthus's arguments were presented as scientific, they nonetheless invoked God as the final cause and promoted God's will and Christian morals as the justification for the elimination of the Poor Laws. The general anger of the working classes toward Malthus and his natural theology (raised to the level of economic science) was best expressed by the political radical William Cobbett, who, in the same general spirit as Marx, said of Malthus: "I have during my life, detested many men; but never any one so much as you. . . . No assemblage of words can give an appropriate designation of you; and, therefore, as being the single word which best suits the character of such a man, I call you *Parson*, which amongst other meanings, includes that of Boroughmonger Tool."[47]

In contrast to these objections to Malthus, Marx strongly defended the scientific character of Adam Smith's economics against the criticisms of theologian and political economist Thomas Chalmers, who considered Smith to have rejected the

Christian view in his close connection to Hume (who was influenced by Epicurus's materialism) and in his concept of unproductive labor, which Chalmers viewed as an attack on God's clergy. In his political economic writings, Marx argued, Chalmers allowed religion and God, complete with "Christian priestly trimmings," to intrude directly into science. "The parsonic element is . . . in evidence not only theoretically but also practically, since this member of the Established Church defends it 'economically' with its 'loaves and fishes' and the whole complex of institutions with which this Church stands or falls."[48]

The Death of Teleology

The materialist conception of nature and the materialist conception of history were for Marx the two indispensable bases of modern science. Human history and natural history ultimately constituted a single historical frame of reference. He therefore consistently advanced evolutionary views against all notions of design by a deity. Life, he contended, had originated in the world in accordance with some kind of spontaneous generation. He argued together with Engels in *The German Ideology* in 1846 that organic existence could not be understood in teleological terms, but involved "the bitterest competition among plants and animals" in which the relation of species to natural conditions was the material cause. And early on he adopted the conception of deep time arising from historical geology.[49]

Marx's admiration for Darwin's evolutionary theory is well known. He was reported as speaking of nothing else for months after the publication of the *Origin of Species*. Upon reading Darwin's work shortly after it appeared Marx wrote to Ferdinand Lasalle: "It is here that, for the first time, 'teleology' in natural science is not only dealt a mortal blow but its rational meaning is empirically explained."[50] His only criticism of Darwin was that by drawing on Malthus for inspiration in developing his theory of natural selection he had inadvertently given credence within the social realm to the Malthusian doctrine, which had espoused Christian morality, natural theology, and bourgeois justifications of the division of class and property. Hence, Marx and Engels sought at all times to separate Darwinian theory from Malthusianism or social Darwinism, while adhering to a materialist/humanist science, seeking to further human freedom.

In place of Malthus's abstract law of population, which was meant to justify class relations, Marx turned increasingly to the new field of anthropology to develop a historical, materialist, and scientific understanding of the development of human populations and societies in all of their aspects. He pointed out that, just as Darwin had referred to the organs developed by species as a kind of "natural technology," the result of natural selection, so too were human tools an extension of the organs of human beings and the product of social evolution. Did not the evolution of the tools of human beings provide, then, an approach to the evolution of human society that required "equal attention? And would not such a history be easier to compile, since, as Vico says, human history differs from natural history in that we have made the former, but not the latter?"[51]

Significantly, at the very time that Darwin was introducing his theory of evolution by natural selection, a second, no less serious, assault on the biblical view of the world was taking place. The year 1859, the date of the publication of Darwin's *Origin*, also marked the beginning of what has been called the "revolution in ethnological time."[52] Although Neanderthal remains had been discovered in 1856, it took time for naturalists to realize exactly what they were. The discovery of prehistoric human remains in Brixham cave near Torquay in southwestern England in 1859 served as conclusive scientific evidence that human beings had existed on earth in great antiquity.[53] This extended the human time line far beyond recorded history, contradicting the view based on the Bible that humanity had existed at most only a few thousand years. Suddenly scientists were faced with evidence that human beings had evolved over a period of time much longer than biblical literalists allowed for the history of Earth. Biologists and geologists closely associated with Darwin, such as John Lubbock and Thomas Huxley, began to consider the question of human evolution, relying in part on what was being revealed of the prehistoric record.

Lubbock built his work on Epicurus/Lucretius's distinction of the stone, bronze, and iron ages. Meanwhile, Lewis Henry Morgan introduced his pioneering work in anthropology, *Ancient Society*, based principally on his studies of the Iroquois—tracing the roots of his own evolutionary perspective to Lucretius.[54] Much of Marx's research for the remainder of his life, after the publication of *Capital*, volume 1, in 1867—even taking precedence over his economics—was devoted to wider ethnological studies as represented by his *Ethnological Notebooks* (1880–82). Marx's approach was built on Morgan's, in the sense of attempting to understand the full development of human productive and familial relations—recognizing that a genuine human anthropology of prehistory was now conceivable. It thus constituted an expansion of science's magisterium at the expense of the magisterium of religion.

Hence, although Marx devoted the greater part of his adult life to developing a critique of the regime of capital as a form of class-based production, this has to be seen as part of a much more fundamental materialist/humanist worldview that arose from his critique of religion. Like Hume, Marx was fond of referring not only to Lucretius but also to the later satirist (and Epicurean) Lucian (c. 120–180) and his *Dialogues of the Gods*, in which, according to Marx, the gods died a second death due to comedy. And just as Hume had turned to Lucretius and Lucian on his deathbed, Marx's response to death, as recounted by Engels, was to quote Epicurus: "Death is not a misfortune for him who dies, but for him who survives."

Indeed, Epicurus, Marx pointed out, argued that "the world must be *disillusioned* and especially freed from fear of gods, for the world is my *friend*." Lucretius had written, "Things come into being without the aid of the gods." Marx added that all human history, including the development of human nature, the formation of new needs, etc., is made by human beings as self-mediating beings of nature, *without the aid of gods*.[55]

6. On the Origin of Darwinism

"The old argument of design in nature, as given by Paley, which formerly seemed to me so conclusive, fails," Charles Darwin noted in his *Autobiography*, "now that the law of natural selection has been discovered. We can no longer argue that, for instance, the beautiful hinge of a bivalve shell must have been made by an intelligent being, like the hinge of a door by man. There seems to be no more design in the variability of organic beings and in the action of natural selection, than in the course which the wind blows. Everything in nature is the result of fixed laws."[1] Darwin's theory of transmutation of species was strictly a materialist conception of nature, dethroning both religious teleology and anthropogenic views. He insisted that the world, in all of its grandeur, be explained in terms of itself. He referred to and adopted Bacon's view that any concept of nature rooted in final causes was "barren, and like a virgin consecrated to God produces nothing."[2] As a result, today's intelligent design advocates attack Darwin and his science.

Darwin's theory of evolution directly challenged design arguments and the prevailing values of Victorian society, as he demonstrated that all species—including humans—were not created directly by a divine being, but rather were the products of natural laws and historical contingency. He established a continuity between human beings and other animals, in bodies, minds, and emotions, that united all animate life by a common set of material relations and evolutionary laws. He scuttled the notion that God created the earth and all its creatures expressly for human beings.

Darwin "elaborated his views on nature and human nature within a larger vision of a world continuously active in the generation of new forms of life and mind. This was materialism, and Darwin knew it; but it was a materialism that humanized nature every bit as much as it naturalized man."[3] Whereas the critique of design can be traced back to antiquity, Darwin's articulation of a materialist evolutionary theory was distinct for its originality, and his materialist account of life constituted an intellectual revolution.

The Emergence of an Evolutionist

Darwin came from a family of Unitarians and freethinkers who embraced the Enlightenment. His father was a medical doctor. His grandfather, Erasmus Darwin, wrote *Zoonomia*—a book that put forward an evolutionary argument about the perpetual transformation of life. Erasmus Darwin's dedication to Reason led him to reject that providence was necessary to ensure the revolution of the earth around the sun.[4]

Early in his upbringing and education, Charles Darwin was thus exposed to the major conflicts between design and materialism. He witnessed the attempts of others to put forward materialist arguments in regard to human development, as well as the censuring of them. He was sent to Edinburgh in 1825 to study medicine. While his interest in this subject waned he started to study natural history and philosophy. He joined the Plinian Society, a student science club that encouraged the study of natural science, the collection of specimens from nature, and the presentation of scholarly work to the Society. At one of the meetings, William A. F. Browne, who nominated Darwin to join the group, levied a critique against Charles Bell's *Anatomy and Philosophy of Expression*. Bell was famous as the author of *The Hand, Its Mechanism and Vital*

Endowments as Envincing Design, the fourth *Bridgewater Treatise*—
a series of eight treatises in natural theology funded by a bequest
from the estate of Francis Henry Egerton, the eighth Earl of
Bridgewater. In the *Anatomy and Physiology of Expression* Bell had
contended that the facial muscles in humans were specifically
designed by God to allow humankind to express unique emotions.
Browne challenged the natural theology argument and presented a
paper on the organization of life, asserting that life was merely a result
of the way the body was organized. He then argued that the "mind,
as far as one individual's senses and consciousness are concerned, is
material." This statement alarmed members of the Society, and his
remarks were stricken from the minutes, after which Browne restricted his inquiries to safer, non-philosophical subjects.[5]

At Edinburgh, Darwin's mentor was Robert Edmond Grant, a
physician and sponge expert. Grant was a materialist and deist who
supported the transmutationist positions of Erasmus Darwin,
Etienne Geoffroy Saint-Hilaire, and Jean-Baptiste Lamarck.
Lamarck was the first person to use the term biology in its modern
sense, and he was the first to develop a cohesive theory of evolution—a theory founded on the inheritance of acquired traits. Grant
encouraged Darwin to read Lamarck's *System of Invertebrate
Animals*. Grant and Darwin often would go for long walks together, collecting specimens. During these walks, Grant would talk
about evolutionary ideas, such as the common origin of the plant
and animal kingdoms.[6]

Lacking interest in becoming a doctor, Darwin was sent by his
father to Cambridge University, at Christ's College, with the
intention that he would take holy orders and become a clergyman
at a country parish. He was amenable to this course, since being
a clergyman would give him the freedom to continue to pursue
his interests in natural history. Here he collected beetles, continued to study Lamarck, and consulted with geologist Adam
Sedgwick, and philosopher, scientist, and theologian William
Whewell, author of the third *Bridgewater Treatise—Astronomy*

and General Physics Considered with Reference to Natural Theology. Darwin's study of natural theology deepened, as he studied William Paley's books: *Evidences of Christianity, Principles of Moral and Political Philosophy,* and *Natural Theology.* Darwin scrutinized Paley's works intensely. These studies gave him "much delight" and he "did not at that time trouble" himself "about Paley's premises." Instead he took them "on trust" given that he "was charmed and convinced by the long line of argumentation."[7] *Natural Theology* was one of the cornerstones of the science program at Cambridge.

Paley catalogued adaptations—some genuine and some far-fetched—of animals to their environment, arguing that these manifestations were the result of contrivance, due to an intelligent and benevolent designer. In recognizing contrivance, humans found proof of the existence, agency, and wisdom of the Deity. Woodpeckers were given long tongues specifically to catch grubs, and the pouches on marsupials were designed to carry young. God even provided anticipatory adaptations, such as the ability of birds to migrate to avoid cold seasons. All adaptations by plants and animals were deemed to be perfect.[8]

Although it is fair to say that transmutation and materialism were in the air, they remained on the fringe of scientific society. Darwin had been exposed to many of the scientific debates. Nevertheless, Darwin not only had professors and mentors that supported the argument from design, but also ones who were open to the question of transmutation of species. From the outset Darwin struggled to understand the complexity of the natural world. When he set sail on what was to be his famous five-year voyage (1831–1836) around the world on the *Beagle,* he did so as a natural theologian still tied to creationist ideas but enough of a free-thinker to be open to questioning them. He was prone to quote the Bible on issues of morality, as if it were "an unanswerable authority." As he later recalled, this voyage was "by far the most important event of my life and has determined my whole career." This adventure radically transformed him and, ultimately, our understanding of the natural world.[9]

On the voyage, Darwin studied "the geology of all the places visited," and he observed and collected specimens of "animals of all classes." He took detailed notes of the vegetation of South America. He discovered fossilized remains of long extinct mammals. Among the unknown fossils were the remains of other species still in existence. Darwin mulled over these finds, trying to understand the causes of extinction, and why both extant and extinct species were found together.[10]

When the *Beagle* stopped at the Galápagos Islands, he took extensive notes and collected specimens to be studied by experts in England. He had been told that species of tortoises varied across islands. He noticed that the same was true of mockingbirds. At the time Darwin failed to notice that several species of apparently dissimilar birds were all in fact finches—this realization did not dawn on him until he was back in England, and proved pivotal to cementing his commitment to evolution. Nonetheless, he pondered the relationship between the species on the different islands and their counterparts on the South American continent. Darwin perceived that the animals on certain islands were often variations of similar species on still other islands. At the time, it was generally accepted among naturalists that this may be due to a slight pliancy as species spread from the point of creation. Yet, Darwin asked "how far a species could be pushed," recognizing that if the divergence from the original stock was sufficiently great, it "would undermine the stability of Species" argument.[11]

While aboard the *Beagle*, Darwin carefully studied Charles Lyell's three-volume collection *Principles of Geology* (1830–33). He marveled at the grand theoretical scheme presented in the books. Lyell's book became one of the most famous scientific textbooks ever written. In it, he attempted to reform the science of geology based on his methodological and substantive doctrine of uniformitarianism, which emphasized that processes observed to operate in the present, such as erosion, if given sufficient time could generate all known geologic formations. Although Lyell's singular focus on uniformitarianism sometimes led him astray, his doctrine was of profound importance for developing geological science, and it inspired Darwin's recognition of how simple natural processes could in the sweep of geologic time yield great transformations.[12]

By the end of the voyage, Darwin's "love of nature was ousting the church." Back in London, he became part of scientific society, becoming friends with Lyell, Robert Owen, John Gould, and Joseph Dalton Hooker, presenting scientific papers, and preparing manuscripts. The fossils and specimens he collected were distributed among scientific experts for classification. Owen determined that the unknown fossils Darwin had found were distant relatives of mammals currently occupying South America. Gould ascertained that the bird specimens from the Galápagos Islands were all finches, differently adapted to the various islands. Further investigation of the mockingbirds, rodents, and tortoises confirmed that the different islands had similar, yet distinct, species.

After reflecting on these observations and other considerations, including the fossil record and the distribution of species in time, in March of 1837, Darwin firmly embraced transmutation, but he had yet to identify the mechanism that drove it.[13] In search of this mechanism, he returned to his *Red Notebook* from his voyage, where he started to raise critical questions about the relationships among species. He also started the first of his transmutation notebooks (*Notebook B*), as he attempted to discern the explanation of organic change, and how humans shared with other animals "one common ancestor."[14]

The working out of a materialist theory of evolution was a slow, arduous process, given that Darwin wanted to define the mechanism of change and to provide extensive details of how it operated. In his *Autobiography* he explained, "It was evident . . . that species gradually become modified; and the subject haunted me. But it was equally evident that neither the action of the surrounding conditions, nor the will of the organisms (especially in the case of plants), could account for the innumerable cases in which organisms of every kind are beautifully adapted to their habits of life."[15] In part, his determination to provide an irrefutable theory stemmed from his awareness of the dangers of presenting a strictly materialist argument, even among the leading scientists of the day, not to mention society in general.

Darwin hid his materialism, as he secretly scribbled away in his notebooks working out his theory. He was well aware of what had happened to the surgeon William Lawrence, a materialist who argued in 1819 that living organisms conformed to higher natural laws than those that could be attributed to inanimate nature. Lawrence had been forced to resign from his position at the College of Surgeons and to recant his views. His offending book, *Lectures on Physiology, Zoology, and the Natural History of Man*, was declared blasphemous by the Court of Chancery and its copyright eliminated. Darwin owned a copy of Lawrence's book, referring to it in his notebooks on transmutation.[16] Denying any "vital principle," Lawrence rooted life in organization of matter and bodily organs, denying the existence of any mental property independent of the brain. For this, he was deemed seditious and immoral.

Nevertheless, Darwin continued to hone his critique of intelligent design. For example, William Whewell, in his natural theology, argued that the length of days was specifically adapted to the amount of sleep humans needed. In this, the whole universe adapted to the needs of humans. Darwin exclaimed in his notebooks that such an anthropocentric conception was an "instance of arrogance!!" He amused himself toying with the idea of transmutation, and taking aim at clerical fears of descent, wrote: "If all men were dead then monkeys make men.—Men makes angels." In this Darwin was developing a non-anthropocentric view of evolution, noting that life was a branching tree and that it was "absurd to talk of one animal being higher than another." Depending on the criteria, different animals could be seen as the highest. As he struggled to discover an organic mechanism, he moved away from Lamarck and rejected unidirectional change.[17]

A Materialist Theory of Evolution

In the fall of 1838, Darwin, with the foundations of natural selection coalescing in his mind, turned to Thomas Malthus's *An Essay on the Principle of Population*, and was struck by the implications of the concept of the struggle for existence. Darwin noted that he was

> well prepared to appreciate the struggle for existence which everywhere goes on from long-continued observation of the habits of animals and plants, [and] it at once struck me that under these circumstances favourable variations would tend to be preserved, and unfavourable ones to be destroyed. The result would be the formation of new species. Here, then, I had at last got a theory by which to work; but I was so anxious to avoid prejudice, that I determined not for some time to write even the briefest sketch of it.[18]

It was in engaging and criticizing design arguments that Darwin sharpened his own evolutionary position leading up to his Malthusian insight. In 1838, Darwin took notes on John Macculloch's *Proofs and Illustrations of the Attributes of God from the Facts and Laws of the Physical Universe, Being the Foundation of Natural and Revealed Religion* (1837)—a work that intended to reconcile theology and natural history. In his notes on Macculloch's work, Darwin played his developing theory against "Providential Design." Many of the standard design contrivances are found in Macculloch. For example, the camel supposedly was specifically designed to have the stomach and feet it has for its intended environment—thus creating a perfect union.[19]

Darwin's cutting notes reveal his increasing radicalism. He wrote that employing a design argument opens up a realm of questions with regard to changes in the natural world and imperfections—and their reflection on the sub-optimality of God's creation. He stated that the design argument had been "exhausted." The notion that certain plants and animals are created by "the *will* of the deity" "is no explanation—*it has not the character of a physical law*," it "is therefore utterly useless." He levied a critique of the *Bridgewater Treatises*, noting that all of them were "simply statements of productiveness" and "laws of adaptation" rooted in declarations of final causes. Invoking Bacon's criticism of final causes he wrote: "Consider these barren Virgins."[20]

Darwin filled a number of notebooks on the topic of transmutation. The range and implications of his materialist position became increasingly more evident to him. In his *M* and *N Notebooks*, he broached the subject of morality as being culturally relative rather than ordained by God. Here he challenged Plato's notion that ideas of good and evil are determined by the preexistence of the soul—exclaiming, "Read monkeys for preexistence." "To avoid stating how far, I believe, in Materialism," he wrote at one point, "say only that emotions, instincts degrees of talent, which are heredetary are so because brain of child resemble, parent stock." In 1838, he studied the expressions of monkeys, ascertaining common, universal characteristics shared by humans and other animals. He even noted that the "love of deity" was an "effect of organization," making an aside to himself, "oh you Materialist!"[21] Here he laid the foundation from which a number of his major works arose in the years to come.

Following his great insights of 1838, Darwin sat on his theory letting it mature. For a while he had considered that "saltationism"—abrupt, sudden changes in lineages—might dominate evolution. However, he soon abandoned this view. Instead, he decided that evolution proceeds in a slow, smooth, gradual manner, reminiscent of Lyell's uniformitarianism. In 1842, shortly before moving to the countryside, he finally prepared a 35-page *Sketch* of his theory of descent, not intended for publication, outlining natural selection via the Malthusian mechanism.

Seeking a sounding board for the direction of this work, on January 11, 1844, Darwin confided to Hooker:

> I am almost convinced (quite contrary to the opinion I started with) that species are not (it is like confessing a murder) immutable. Heaven forfend me from Lamarck nonsense of a "tendency to progression," "adaptations from the slow willing of animals," & c.! But the conclusions I am led to are not widely different from his; though the means of change are wholly so. I think I have found out (here's presumption!) the simple way by which species became exquisitely adapted to various ends.[22]

Hooker encouraged Darwin in his pursuits, and Darwin expanded his *Sketch* into a 230-page *Essay*. But he still did not pursue making public his theory. However, he left instructions to his wife, Emma, to publish the *Essay* in case of his own death to ensure that his discoveries would not be lost.

Darwin was a respected scientist. He had clerical and scientific friends. He remained anxious regarding the risk of putting forward his theory. Shortly after he completed his *Essay*, the anonymous *Vestiges of the Natural History of Creation* (later revealed to have been written by Robert Chambers) was published. It presented an argument regarding the creation of life and the universe by natural laws, and argued for the interconnectedness of all life. Although promoting transmutation, *Vestiges* was not of the high-intellectual caliber of Darwin's work, lacking a firm scientific foundation. Darwin thought that the geology and zoology within the book was inadequate. Nonetheless, initially he worried that he may have been partially scooped on his own discoveries. However, he soon recognized that his own ideas remained distinct and much more rigorously developed. *Vestiges* shocked British society. The book went through numerous printings, each time cleaning up mistakes and improving its argument. Darwin's old professor, Sedgwick, a devoted natural theologian, attacked the book and its notion of spontaneous generation, insisting that God operated through creative power rather than transmutation.[23]

Observing the uproar over *Vestiges*, Darwin became convinced that his argument must be impregnable. He recognized that he still had to work out some issues. He explained that one issue overlooked was "the tendency in organic beings descended from the same stock to diverge in character as they become modified," but he realized that "the solution, as I believe, is that the modified offspring of all dominant and increasing forms tend to become adapted to many and highly diversified places in the economy of nature."[24]

In 1856, Lyell, although skeptical of Darwin's views, encouraged Darwin to prepare a preliminary paper outlining his theory for the record. In an effort to provide a complete account, Darwin started a manuscript—entitled *Natural Selection*—originally intended to be three to four times the size of what later appeared as *On the Origins of Species*. He had Thomas Henry Huxley, Hooker, and John Lubbock read selections from this manuscript. In 1857, he prepared a summary of his argument and mailed it to the Harvard botanist Asa Gray.[25]

Ever meticulous in his work, Darwin had been accumulating facts and details to demonstrate his own theory of evolution for twenty years. But his long delay was to come to an end. He was not the only one trying to articulate a materialist theory of evolution. Naturalist Alfred Russel Wallace set out on his expeditions to the Amazon of South America (1848–52) and the Malay Archipelago (1854–62) already believing in transmutation. During his eastern travels, Wallace published an article, "On the Law Which Has Regulated the Introduction of New Species," in 1855 in the *Annals*. Lyell, fearing that Darwin might be scooped, told Darwin about the article and continued to urge him to publish his work to establish precedence. On May 1, 1857, Darwin wrote to Wallace noting that the two of them had been thinking very much alike in regard to the transmutation issue.[26] Both Darwin and Wallace were moving toward revolutionizing scientific thought on the origin of species. Darwin already knew the mechanism, but labored to accumulate a mountain of facts to convince Victorian society. Wallace, although long convinced of transmutation, still struggled to discover the mechanism.

In 1858, Wallace sent Darwin a new paper he was preparing, "On the Tendency of Varieties to Depart Indefinitely from the Original Type; Instability of Varieties Supposed to Prove the Permanent Distinctness of Species." In it, Wallace presented an evolutionary theory based on divergence and the struggle for survival. Darwin recognized that Wallace had seized upon the same materialist mechanism. Although he did not want to be seen as taking advantage of someone else's work, he did not want his own breakthrough discounted. He consulted Lyell and Hooker, who convinced him that he should not simply step aside and give all the credit to Wallace. Lyell and Hooker orchestrated the famous delicate arrangement where an "abbreviated abstract of his 230-page essay from 1844," the letter that Darwin wrote in 1857 to Asa Gray in which he had outlined natural selection, and Wallace's 1858 paper were all read at the Linnean Society meeting on July 1, 1858, and subsequently published in the *Journal of the Proceedings of the Linnean Society of London*, August of that year.[27] Darwin and Wallace were thus presented as co-discovers of natural selection, and each was recognized for reaching his insight independently.

At this point, Darwin worked feverishly to prepare an abstract of *Natural Selection*, in order to present his theory of natural selection in full for a general audience. *On the Origin of Species By Means of Natural Selection, or the Preservation of Favoured Races in the Struggle for Life* was published in November 1859, and Darwin noted that its success was in part due to being a much shorter book than what he had been preparing with *Natural Selection*.[28] With the publication of *Origin of Species*, the dominance of teleological arguments in the treatment of nature, and the class and religious bases of these, began to collapse. Twenty years after he had made one of the most profound discoveries in the history of science, Darwin here finally presented his strictly material-

ist theory of evolution by natural selection. He had at last over-
thrown the hold of natural theology. Materialism triumphed over
teleological views of nature, as it became evident that the very
"divine contrivances" raised by natural theologians in support of
their position were better understood when examined through a
non-theistic, naturalistic lens.[29]

In *Origin of Species*, Darwin accomplished two major feats.
First, he laid out some of the vast evidence he had accumulated
over the preceding two decades that demonstrated the reality of
evolution—that species change over time and are connected to one
another through ancestry. Although the idea of transmutation was
not new, it had never before been presented with such a rigorous
array of evidence in its favor. Darwin's case for the *fact* of evolution
was so strong that despite some initial resistance many in the scientific community quickly accepted it. Darwin's second major task
was to present his *theory* of what drove the evolution of species:
natural selection. This was the most controversial part of his book,
and it remained highly contested well into the twentieth century,
until advances in genetics, biogeography, and paleontology finally
sealed Darwin's case by midcentury.

Darwin noted that natural selection operating on variation
among individuals served as the mechanism through which transmutation occurred, differentiating his position from prior transmutation theories. He explained that organisms tend to produce more
offspring than can survive. The individuals vary from one another
in their traits, and these traits are at least in part heritable (although
Darwin did not know the specific laws of heredity which had to
await the rise of genetics). There is a struggle for survival and those
best fitted via innate characteristics to the environment in which
they reside have a higher survival and reproductive success rate,
thus passing on part of their advantageous traits to their own offspring. The accumulation of such favorable traits results in descent
with modification—the evolution of species, which over sufficient
time can lead to dramatic transformation of a lineage. In a startling

metaphor, he wrote of natural selection: "The face of Nature may be compared to a yielding surface, with ten thousand sharp wedges packed close together and driven inwards by incessant blows, sometimes one wedge being struck, and then another with greater force."[30]

In *Origin of Species*, Darwin avoided the word *evolution*, given that it was then commonly tied to teleological notions of unfolding and to progressive development. In fact, he only used the word *evolved* once in the first edition of the book, in the final sentence: "There is grandeur in this view of life . . . whilst this planet has gone cycling on according to the fixed law of gravity, from so simple a beginning endless forms most beautiful and most wonderful have been, and are being, evolved."

Likewise Darwin as a consistent materialist avoided the notion that evolution always led to higher forms ("never use the words higher and lower" he had written in the margins of his copy of the *Vestiges of the Natural History of Creation*). But in his attempt to reassure readers at the very end of his book (in the second to last paragraph) he observed, in violation of his own precepts: "As natural selection works solely by and for the good of each being, all corporeal and mental endowments will tend to progress towards perfection."[31]

Darwin's characteristic brilliance can be seen in his ability to explain how natural selection led to the adaptation of organisms to their environments, while excluding all final causes. The conditions of a local environment change, so which organisms are best able to survive also changes with time. There is no inherent superiority/inferiority in species, determining that they are predestined to survive. In this, Darwin provided a strictly materialist conception of nature, expanding the magisterium of science to the entire physical world.

Darwin versus Intelligent Design

One of the most important books that influenced Darwin, by his own account, was John Herschel's *A Preliminary Discourse on the Study of Natural Philosophy* (1831). Herschel was one of the leading British scientists of the age, known for his work in astronomy, geography, and scientific method. *Discourse on the Study of Natural Philosophy* provided a model of the interplay between observation (experience) and theory. Herschel insisted that laws govern nature, but these laws are often difficult to determine. Attempting to understand these laws was the ultimate goal of natural philosophy. In this, researchers could discover the fundamental truth of nature unified in a single explanation. In the *Origin of Species*, Darwin sought to adhere to Herschel's scientific method and argument. Darwin noted in the opening pages that he was attempting "to throw some light on the origin of species—that mystery of mysteries, as it has been called by one of our greatest philosophers." The great natural philosopher who spoke of the "mystery of mysteries" was Herschel.[32]

But the *Origin of Species* called into question what had been known about the living world, arguing that species were not immutable and that there was a continuum of life, produced by naturalistic causes. In this, Darwin's theory of evolution via natural selection challenged established thought, including some of the most prominent scientists who had influenced him. Darwin sent a copy of the *Origin of Species* to Herschel as a token of his admiration. To his great consternation, Darwin subsequently heard that Herschel had "called natural selection the law of higgledy-piggledy."[33]

Herschel commented publicly on natural selection in the 1861 edition of his *Physical Geography of the Globe*, which he sent to Darwin. In a footnote, Herschel questioned Darwin's theory, given that it suggested that variations could occur in all directions, even ones that did not directly help species adjust to their dynamic environments. Such variations, in Darwin's theory, "gave no indication of the Creator's foresight." In sharp opposition to this Herschel wrote:

> We can no more accept the principle of arbitrary and casual variation and natural selection as a sufficient account, *per se*, of the past and present organic world, than we can receive the Laputan method of composing books (pushed *à l'outrance*) as a sufficient one of Shakespeare and the *Principia*. Equally in either case, an intelligence, guided by a purpose, must be continually in action to bias the directions of the steps of change—to regulate their amount—to limit their divergence—and to continue them in a definite course. We do not believe that Mr. Darwin means to deny the necessity of such intelligent direction. But it does not, so far as we can see, enter into the formula of his law; and without it we are unable to conceive how the law can have led to the results.

Although critical and concerned with the implications of Darwin's theory—contending that evolutionary changes required "intelligent direction" and that a special stipulation needed to be made for humans—Herschel did note enigmatically in the end: "We are far from disposed to repudiate the view taken of this mysterious subject in Mr. Darwin's work."[34]

Responding to Herschel on May 23, 1861—thanking him for sending *Physical Geography* and commenting on the reference to natural selection—Darwin took issue with what he called the notion of "intelligent Design," introducing this term for the first time in its modern sense:

> The point which you raise on intelligent Design has perplexed me beyond measure; & has been ably discussed by Prof. Asa Gray, with whom I have had much correspondence on the subject. . . . One cannot look at this Universe with all living productions & man without believing that all has been intelligently designed; yet when I look to each individual organism, I can see no evidence of this. For I am not prepared to admit that God designed the feathers in the tail of the rock-pigeon to vary

in a highly peculiar manner in order that man might select such variations
& make a Fan-tail; & if this be not admitted...then I cannot see design in
the variations of structure in animals in a state of nature,—those variations which were useful to the animal being preserved & those useless or
injurious being destroyed.[35]

Darwin insisted that his theory of natural selection required no
recourse to "intelligent direction" or a "Higher power." Still reflecting upon Herschel's contention that "the higher law of providential arrangement" should always be noted and Asa Gray's insistence on divine guidance of variation, Darwin confided to Lyell on
August 1, 1861:

> But astronomers do not state that God directs the course of each comet
> & planet.—The view that each variation has been providentially
> arranged seems to me to make natural selection entirely superfluous, &
> indeed takes whole case of appearance of new species out of the range of
> science. . . . I wonder whether Herschel would say that you ought always
> to give the higher providential Law, & declare that God had ordered all
> certain changes of level that certain mountains should arise.— I must
> think that such views of Asa Gray & Herschel merely show that the sub-
> ject in their minds is in Comte's theological stage of science [the first of
> three stages in the development of knowledge].[36]

Darwin clearly sensed the ramifications of his materialist theory
with its tacit recognition of the vast indifference of nature to human
affairs. Calling into question nature as evidence of God, Darwin
wrote to Asa Gray on May 22, 1860:

> With respect to the theological view of the question: This is always
> painful to me. I am bewildered. I had no intention to write atheistically.
> But I own that I cannot see as plainly as others do, and as I should wish
> to do, evidence of design and beneficence on all sides of us. There seems
> to me too much misery in the world. I cannot persuade myself that a
> beneficent and omnipotent God would have designedly created the
> Ichneumonidae with the express intention of their feeding within the liv-

ing bodies of Caterpillars, or that a cat should play with mice. Not believing this, I see no necessity in the belief that the eye was expressly
designed. On the other hand, I cannot anyhow be contented to view this
wonderful universe, and especially the nature of man, and to conclude
that everything is the result of brute force. I am inclined to look at everything as resulting from designed laws, with the details, whether good or
bad, left to the working out of what we may call chance.[37]

Darwin remained steadfast in his opposition to intelligent
design within the physical world, including all living beings. In a
letter to Julia Wedgwood, on July 11, 1861, he distinguished his
position from advocates of design: "The mind refuses to look at
this universe, being what it is, without having been designed; yet,
where one would most expect design, viz. in the structure of a sen-
tient being, the more I think on the subject, the less I can see proof
of design. Asa Gray and some others look at each variation, or at
least at each beneficial variation (which A. Gray would compare
with the rain drops which do not fall on the sea, but on to the land
to fertilize it) as having been providentially designed."[38]

The Descent of Man

Following *Origin of Species*, each of his major subsequent works
focused on a specific substantive issue while simultaneously developing the methodology of his evolutionary theory.[39] Lyell and Gray
contended that variation was directed by providential influence,
rather than "pure chance." Darwin's *The Variation of Animals and
Plants under Domestication* (1868) attempted to lay these critiques
to rest by detailing how breeders specifically selected traits from an
array of variations. He argued that the traits selected by breeders
were not preordained for their express purpose: "No shadow of
reason can be assigned for the belief that variations, alike in nature
and the result of the same general laws, which have been the
groundwork through natural selection of the formation of the most

perfectly adapted animals in the world, man included, were intentionally and specifically guided."[40]

Darwin's work on plants was also an extension of his materialistic theory of evolution, as well as a critique of design. Design proponents proposed that beauty abounded in nature expressly for the delight of humans. Darwin opposed this view, and prepared an extensive manuscript *On the Various Contrivances by which British and Foreign Orchids Are Fertilised by Insects* (1862) as a "flank movement" against design advocates. He argued that natural selection could explain plant morphology and physiology. For example, he contended that the ornate ridges and horns of orchids were "adaptations to facilitate reproduction" by their interaction with insects.[41]

It was in *The Descent of Man* (1871) that Darwin first took up the critical issue of human origins. When the *Origin of Species* was published in 1859, there was a concurrent challenge to the biblical view of the world, which, as we noted in chapter 5, has been called the "revolution in ethnological time."[42] In 1859 prehistoric human remains were discovered in Brixham Cave near Torquay in southwestern England, providing convincing evidence for the first time (though Neanderthal remains had been found earlier near Düsseldorf) that human beings had existed on earth in great antiquity. The human time line was extended far beyond recorded history, undermining the view based on the Bible, sacrosanct up until that time, that humanity had existed at most only a few thousand years. Here was scientific evidence that human beings had evolved over a period of a million years or more. Biologists and geologists closely associated with Darwin, such as Lubbock and Huxley, began to consider the question of human evolution, relying in part on what was being revealed of the prehistoric record.[43] These works extended the discussion of human society to take into account the longer conception of ethnological time and to provide a general theory of human social development, rooting human history in changing material conditions.[44]

Darwin entered into this discussion, building on his theory of natural selection. Whereas some had tried to reconcile the argument in the *Origin of Species* with religious views, *The Descent of Man* shattered such positions, given its explicit materialism. Instead of invoking the hand of God as an explanation of human origins, Darwin proposed a materialist approach as the basis for investigation. In support of science, he noted, "Ignorance more frequently begets confidence than does knowledge: it is those who know little, and not those who know much, who so positively assert that this or that problem will never be solved by science." He united humans and other creatures via evolution by common descent, arguing that the same materialistic forces influenced the historical development of all life. He presented the natural, material origins of society, morality, and religion, banishing the divine hand from the world. His discussion of sexual selection—somewhat distinct from natural selection—was in part a critique of design arguments. For instance, the tails of peacocks were often seen as "an expression of divine aesthetics—beauty as an end in itself and incapable of natural explanation." In case after case, Darwin sought to explain the evolutionary emergence and development of beauty, as a result of sexual selection, a strictly materialist process, based on the sexual displays of animals.[45]

For Darwin, all animate life was united by a common set of material relations and evolutionary laws. In *The Expression of the Emotions in Man and Animals* (1872), Darwin undermined the traditional anthropocentric interpretation that divided animals from human beings. He destroyed the notion that God created the earth and all of its creatures for humankind. As part of his evolutionary theory, he sought to show the continuity of species and that "humans are not a separate divinely created species."[46]

This work, in part, returned to an issue that Darwin witnessed in the Plinian Society, when Browne had boldly challenged the argument of Bell's *Anatomy and Philosophy of Expression*, and that he had written about in his notebooks on animal expressions. Bell contended that the Creator designed the muscles in the human face specifically for the display of human emotions, and that they could not be found in the rest of the animal kingdom. Darwin challenged Bell's design argument throughout *Expression of the Emotions*, illuminating the continuity in the minds and emotions of human beings and animals. Human expressions and muscles are not unique. Rather, they are the result of evolution. Furthermore, human "expressions are innate" and "are the product of evolution."[47] Here Darwin extended his theory that all species descend from common progenitors, banishing design arguments from claiming that human beings were the exception to natural selection.

Darwin and Religion

On September 28, 1881, Darwin hosted a group of freethinkers for dinner at his Down House. Edward Aveling (who was later to become the common-law husband of Marx's daughter Eleanor) and the leading German scientific materialist Ludwig Büchner were among those in attendance. In the discussion that followed, Darwin admitted that he had finally given up completely on Christianity at forty years of age, but that he was agnostic on the issue of God. Religion remained an issue, whether in the open or as an undercurrent, throughout Darwin's life. He often contemplated the ramifications of his theory, as well as the role of science in society, while generally refusing to pronounce on religion. But it is his commitment to science that shines through his own reflections.

As Darwin focused on the material world and the fixed laws of nature, he recognized that the "false history of the world" as presented in the Old Testament and other sacred books could not be trusted. He rejected divine revelation and contended that it was irrational to believe in miracles. In his *Autobiography*, he noted, "I gradually came to disbelieve in Christianity as a divine revelation." He found evidence in scripture lacking, and finally "disbelief crept over me at a very slow rate, but was at last complete. The rate was so slow that I felt no distress, and have never since doubted even for a single second that my conclusion was correct." He referred to the notion of eternal punishment for non-believers as "a damnable doctrine."[48] Darwin rejected "the existence of an intelligent first cause" as he put his materialist theory of evolution against religious justifications for God. He reflected upon the "argument for the existence of an intelligent God" that was based on "deep inward conviction and feelings" that people experience. He explained that not all people experience the same

> inward conviction of the existence of one God. . . . Therefore I cannot see that such inward convictions and feelings are of any weight as evidence of what really exists. The state of mind which grand scenes formerly excited in me, and which was intimately connected with a belief in God, did not essentially differ from that which is often called the sense of sublimity; and however difficult it may be to explain the genesis of this sense, it can hardly be advanced as an argument for the existence of God, any more than the powerful though vague and similar feelings excited by music.[49]

In *Descent of Man*, Darwin engaged religion as part of the evolutionary development of human societies. He argued that religion was not innate, given that "there is no evidence that man was aboriginally endowed with the ennobling belief in the existence of an Omnipotent God." He explained that some cultures don't have a concept of God or gods.[50] He argued that belief in God was a product of culture:

> I am aware that the assumed instinctive belief in God has been used by many persons as an argument for His existence. But this is a rash argument, as we should thus be compelled to believe in the existence of many cruel and malignant spirits, possessing only a little more power than man; for the belief in them is far more general than of a beneficent Deity. The idea of a universal and beneficent Creator of the universe does not seem to arise in the mind of man, until he has been elevated by long-continued culture.[51]

Yet, although Darwin remained committed to a materialist-scientific approach to the world, he shied away from outright attacks on religion. In this respect, Darwin's approach was widely divergent from the radical, science activists of his day, such as Aveling, who was an anatomy lecturer, and spoke "about Christian hypocrisy and the curtailment of civil liberties" in Britain. Aveling was at the forefront of protests over Charles Bradlaugh, a militant atheist, being denied his elected seat in the House of Commons. In 1880, Aveling asked permission to dedicate his book *The Student's Darwin* (1881) to Darwin. The latter declined the offer, noting that

> though I am a strong advocate for free thought on all subjects, yet it appears to me (whether rightly or wrongly) that direct arguments against christianity & theism produce hardly any effect on the public; & freedom of thought is best promoted by the gradual illumination of men's minds, which follow[s] from the advance of science. It has, therefore, been always my object to avoid writing on religion, & I have confined myself to science. I may, however, have been unduly biased by the pain which it would give some members of my family, if I aided in any way direct attacks on religion.[52]

Darwin depended upon scientific inquiry to expand the magisterium of science, hoping that through careful detailed investigations of the world, an improved understanding of nature would be possible.

Darwinian Revolution

Darwin's materialist account of evolution by means of natural selection, or descent with modification, was a sophisticated theory backed at every point by copious data—one that has inspired scientists ever since. In this respect, it is by no means an exaggeration to speak of the "Darwinian Revolution." As the geneticist Theodosius Dobzhansky declared, "Nothing in biology makes sense except in the light of evolution."[53] Evolutionary (and proto-evolutionary) thought did not of course begin with Darwin. But the Darwinian theory of natural selection represented a qualitative breakthrough, and evolutionary theory has developed substantially since Darwin's time.[54] It initiated a revolution in how we understand nature, adhering to an approach that allowed for a rational inquiry into a dynamic world, while demanding that we explain the world in terms of itself. In this it banished God from the physical world. Darwin laid the foundation on which the biological sciences have built since his time.

Hence, it is only ignorance that allows intelligent design proponent Dembski to declare that "no significant details [in evolutionary theory] have been added since the time of Darwin (and, one can argue, none has been added even since the time of Empedocles and Epicurus two thousand years earlier)."[55] Likewise, Wiker contends that, "We see in this passage [from Epicurus/Lucretius] all the fundamentals of Darwin's account. . . . Darwin followed Epicurean materialism in eliminating species distinctions, and his account of natural selection was also an expansion of that which occurs in Lucretius."[56] The strategy of such intelligent design proponents is to claim that the entire framework of Darwinian evolutionary theory was already laid in ancient times in the work of Epicurus and all that Darwin added was questionable evidence. Such views are absurd, despite the genius often displayed by the ancient Greek materialists, since Darwin provided the first consistent *theory of natural selection*, as well as the evidence to support it.

Ernst Mayr, one of the most renowned evolutionary theorists of the twentieth century, notes in his extensive history of biological thought that although the Epicureans had many important insights the attribution to them of the discovery of natural selection and a developed evolutionary theory is a misinterpretation. Furthermore, in contrast to the state of scientific thought at the time, Darwin's evolutionary mechanism of change was revolutionary. "The fixed, essentialistic species was the fortress to be stormed and destroyed; once this had been accomplished, evolutionary thinking rushed through the breach like a flood through a break in a dike."[57] Darwin provided a groundbreaking theory that went beyond what had come before and sounded the death knell to teleological conceptions of the world.

7. Freud and the Illusions of Religion

Like Karl Marx, the founder of historical materialism, Sigmund Freud, the founder of psychoanalysis, was heir to the critique of religion emanating from the materialist tradition in general and Feuerbach in particular. As German Catholic theologian Hans Küng observed in the first sentence of his *Freud and the Problem of God*: "The grandfather of Marxist atheism and of Freudian atheism is Ludwig Feuerbach, who was first a theologian, then a Hegelian, and finally an atheistic philosopher." "Among all the philosophers," the young Freud wrote, "I worship and admire this man [Feuerbach] the most."[1]

In the later, scientific-materialist phase of his development, Feuerbach had given credence to the mechanistic materialism underlying the work of such nineteenth-century scientists as Jakob Moleschott, Carl Vogt, and Ludwig Büchner.[2] One of Freud's youthful "idols," associated with the same tradition, was the physician and physicist Hermann Helmholtz, a co-discoverer of the conservation of energy. Freud, who early in life was interested in medicine, neurology, and other biology topics, became an exponent of the mechanistic physiology as propounded in particular by Helmholtz, Émil Du Bois-Reymond, Ernst Brücke, and others. He was a strong admirer of British materialists and evolutionary thinkers such as Charles Darwin, John Tyndall, and Thomas Huxley.

As Freud shifted the focus of his studies to the mind, he based the foundations of his psychoanalysis, or psychology of the unconscious, on rigorous materialist and biological principles, concentrating on the interaction between physiological and psychological factors.[3] This debt to biological research and theory has often been obscured due to Freud's later efforts to present his psychoanalytic theory as having been derived independently as a "pure psychology," arising from his fabled "'self-analytic' path of discovery." Nevertheless, the materialist and physiological foundations of psychoanalysis were too essential to the overall structure of Freud's thought to be removed. Indeed, the biological influence on his thinking was so strong as to lead one noted biographer to refer to him as a "crypto-biologist."[4] So consistent and thoroughgoing was Freud's materialism that, according to Küng, it could be traced back to the tradition initiated in antiquity by Democritus, Epicurus, and Lucretius, and to Enlightenment materialists such as La Mettrie and Holbach. Similarly, Erich Fromm observed that "Epicurus' [psychological] theory resembles Freud's in many ways." Freud's deep commitment to materialist and evolutionary theory, no doubt encouraged his follower Ernest Jones to bestow on him the not entirely inappropriate title of "Darwin of the mind."[5]

It is customary to see Freud as having "grown up," as Peter Gay has written, "with no religious instruction at home," and as an atheist even prior to his university years. Thus Freud's early follower and biographer, Ernest Jones, observed that Freud "grew up devoid of any belief in a God or Immortality." Yet, we know that Freud's father had been educated as an Orthodox Jew and never converted to Christianity or completely abandoned his faith, while Freud's mother also retained faith in a deity. Freud was thoroughly acquainted with all Jewish customs and festivals. As a "seven-year-old boy he was intensely interested in the Philippson Bible, a bilingual edition (Hebrew and German) of the Old Testament prepared by the Leipzig rabbi Philippson. This was the standard edition of the Holy Scripture read by emancipated Jews in the nineteenth century." Freud's father gave him an inscribed copy of a Bible on his thirty-fifth birthday.[6] Although he later frequently characterized himself as a "godless Jew" or "infidel Jew," Freud never entirely threw off his Jewish cultural heritage.[7] His lifelong fascination with the character of Moses was revealed both in his article on "The Moses of Michelangelo" in 1914 and in his final work on *Moses and Monotheism* published just before his death in 1939.

Nevertheless, Freud's materialist bent and his commitment to science, from at least his student days on, made him, in the words of Gay, "a convinced, consistent, aggressive atheist."[8] Ernest Jones wrote, with an implication that Freud would no doubt have approved, "A skeptical friend of mine once remarked that the only argument he knew in [favor of monotheism] was a purely arithmetical one: monotheism was nearer the truth because one is nearer to zero than three or five."[9]

Beginning in 1907 in his article "Obsessive Actions and Religious Practices" Freud showed interest in the parallels between psychological neuroses and religion, which eventually led him to his theory of the origin of religion, and his larger critique of religion. In works such as *Totem and Taboo* (1913) and *Moses and Monotheism* Freud sought to provide a theory of the psychogenesis of religion, drawing on Darwinian and Lamarckian evolutionary notions, which, although ultimately unsuccessful, raised pregnant questions about the role of religion in Western civilization.[10] Feuerbach in his famous projection theory had argued that human beings made God in their own image. Freud went still further and sought to demonstrate that human beings made God in the image of the primal father, who had been murdered (and eaten) by his sons in a prehistoric acting out of the Oedipus Complex (sexual desire for the mother and hence rivalry with the father), and that this had been encoded in the species and carried forward by means of a process of Lamarckian evolution (involving the inheritance of acquired characteristics).

Freud's critique of intelligent design, however, was not limited to his psychogenesis of religion. He went beyond his specific theory of the genesis of God as a psychological construction to provide—in works such as *The Future of an Illusion* (1927) and in the final lecture of his *New Introductory Lectures on Psychoanalysis* (1933)—a general critique of religion as an illusory competitor to materialist science.

Along with the notoriety arising from his treatment of neuroses as rooted ultimately in repression of sexual energy, this critique of religion has earned Freud the ire of today's intelligent design proponents. Together with Darwin and Marx, Freud is thus the third member of the unholy trinity continually referred to by fellows of the Discovery Institute's Center for Science and Culture as personifying a godless materialism, emanating from ancient Epicurean roots. Intelligent design's wedge theorist, Phillip E. Johnson, sees Freud along with Darwin and Marx as constituting the "three giants of materialism" and irreligion.[11] For intelligent design proponents Donald De Marco and Benjamin Wiker, Freud is one of the chief modern "architects of the Culture of Death": "Freud, in effect, reduced the world of man and all his distinctly human operations to mere fodder for scientific materialism. . . . Freud entered into a 'Satanic pact' and . . . psychoanalysis was its result. Soon after the pact . . . Freud wrote *The Interpretation of Dreams* . . . which he always regarded as his masterpiece."[12] Freud is naturally accused of having built an analysis of human development around sexual pleasure, leading to accusations that, like Darwin and Marx, he contributed to the destruction of the divine meaning of life.[13] He even removed the gods from dreams themselves, making them utterly material. But the antagonism of intelligent design proponents toward Freud goes much deeper than this, and can be seen as a response to his well-known materialism and atheism and his critique of religion as an "illusion."

By the late 1920s Freud had concluded that religious fundamentalism was increasingly taking on the role of aggressor, challenging science in its own domain. He argued in his *Future of an Illusion* and his *New Introductory Lectures on Psychoanalysis* that science could not afford to be put on the defensive in this respect. Freud was disturbed by the famous 1925 Scopes Trial in Dayton, Tennessee, in which John Scopes, a high school science teacher, was charged with illegally teaching evolution. The trial resulted in the conviction of Scopes (later overturned on a technicality). For Freud "the Americans who instituted the 'monkey trial' at Dayton have alone shown themselves to be consistent" in recognizing the absolute conflict between science and creationism. Progress in knowledge required that traditional religion be seen not as a competitor with science in the realm of truth, but as an illusion.[14]

In order to understand Freud's critique of religion, and especially his psychogenesis of religion, it is necessary to recognize, as he explained in *Civilization and Its Discontents*, that it was aimed at "what the common man understands by his religion. . . . The common man cannot imagine this Providence otherwise than in the figure of an enormously exalted father." Indeed, for Freud this was the "only religion which ought to bear that name." Freud therefore rejected what he called "a series of pitiful rearguard actions," such as deism, which all too often attempted to support the illusion of religion and at the same time replaced God with "an impersonal, shadowy, and abstract principle." As Freud noted in *The Future of an Illusion*, a deism or philosophical theism that is confined "to a belief in a higher spiritual being, whose qualities are indefinable and whose purpose cannot be discerned," may be "proof against the challenge of science," but it will lose its "hold on human interest."[15]

In his university days Freud was influenced for a time by the philosophical theism of Franz Brentano, an ex-priest and professor of philosophy at the University of Vienna. Brentano was a distinguished exponent of Aristotle, a pioneer in empirical psychology, a strong proponent of the teleological argument for the existence of God, and a no-holds-barred critic of Darwinian evolution. Freud considered Brentano a "genius" and a "sharp dialectician" and took no less than five courses from him. Viennese students were enamored with Brentano's lectures in psychology and flocked to his lectures on the existence of God in ever larger numbers, requiring that he be moved into the biggest lecture hall at the university.

In *On the Existence of God: Lectures Given at the Universities of Würzburg and Vienna (1868–1891)* Brentano presented a number of proofs of the existence of God that he considered valid, the first and last of which were, respectively, "the teleological proof, from the rational order in nature," and "the psychological proof, from the nature of the human soul."[16] The "teleological proof," which most occupied Brentano, argued the traditional position of natural theology that the appearance of design in nature led upon examination of the evidence to the proof of the reality of design, and from there to a designer. Thus, for example, he purported to demonstrate that it was design that necessarily accounted for vision in living things. But more important than traditional design notions, in Brentano's teleological proof, was his extensive critique of Darwinian evolution.

Brentano supported the notion of evolution insofar as this left room for teleological principles (and hence God), but he passionately denied Darwinian natural selection, as the one view most opposed to a theological outlook. His lectures carried the headings "Darwin's Explanation Is Not a Safe Assumption," "The Darwinian Hypothesis Is Highly Improbable," and "The Impossibility of the Darwinian Hypothesis." He used a battery of arguments, such as the criticism of blind necessity/chance; the variability of inheritance from seeds, supposedly contravening Darwinian mechanism; the physicist William Thomson's (Lord Kelvin's) calculations (subsequently proven wrong) that Earth was too young for Darwinian natural selection to account for the development of life; the contradictions between Lamarckian and Darwinian evolution, etc. Brentano's whole case was that evolution could not have occurred, contrary to Darwin, without the help of teleological principles. He repeatedly locked horns with Hume's *Dialogues Concerning Natural Religion*. Brentano was particularly critical of all materialist ideas and thus frequently targeted Frederick Albert Lange's *History of Materialism*. He criticized the atomism derived from the ancient Greek materialists. Brentano concluded his teleological proof by arguing that "the apparent teleology in the world, which obliged us to assume a world-ordering intelligence, obliged us also to assume a creative intelligence which has produced the matter itself, which was to be ordered, out of nothing."

More important, perhaps, than Brentano's "teleological proof" of the existence of God, for Freud, was his "psychological proof," in which Brentano again took on evolutionary views. His "psychological proof" took the form of an insistence that the soul, from which psychological attributes derived, was separate from the body. This led to the view that "what thinks in us is not something corporeal, that it must rather be held to be something spiritual." This "spiritual substance," he argued, "must have been produced by a consciously operating creative principle"—indeed God.

In an ancillary argument challenging materialist views related to psychology, which was no doubt to have a lasting impact on Freud, Brentano insisted that Darwinian thinkers themselves pointed to such a separation between body and psyche (soul). Thus, though "a majority of . . . Darwinians" (referring to the Lamarckian-Darwinians of his day) claimed the inheritance of acquired characteristics was "possible," this was viewed as applying only to physiological characteristics, implying that the psyche (soul) was distinct from the body after all. "One could point," Brentano argued, "to the fact," agreed to by all Darwinians, "that no acquired idea, no acquired knowledge, is ever passed on from parents to children." But this "would have to be called strange," he added, "if the soul were a bodily organ, due to the contrast with other inheritable characteristics." Indeed, "from the standpoint of all Darwinian factions," he observed, "the universal fact that there is no inheritance of sense perceptions, cognitive concepts and knowledge" is strictly adhered to. And this "is the opposite of what one would have to expect on the assumption that the soul is corporeal."

Brentano saw this as a deep contradiction in the view of the Darwinians of his day, who, with few exceptions, believed that acquired corporeal characteristics could be inherited (adopting Lamarckian assumptions in this regard), conceived the soul as corporeal, yet universally denied the inheritance of acquired characteristics where higher consciousness was concerned.

There is no doubt that Brentano was confused here about the underlying logic of Darwinism, since it is one of the defining characteristics of Darwinism as such that it denies the inheritance of acquired traits (i.e., Larmarckianism). Hence, Darwinism itself was not in any way challenged by Brentano's critique related to the inheritance of acquired characteristics, which only applied to the Lamarckianized Darwinism then popular.

Still, there is reason to believe that Freud was deeply affected by Brentano's argument on the inheritance of acquired characteristics and the inconsistencies of Darwinism (actually Lamarckian-Darwinism) in this respect. The strong effect that Brentano's argument in his psychological proof likely exerted on Freud's thinking no doubt helps to explain his subsequent radical departure from the views common among Lamarckian-Darwinian biologists of his day, causing him to move not toward but away from Darwin. Thus Freud adopted a strong Lamarckianism, one that supported inheritance of acquired characteristics not only of a corporeal nature (as Lamarckian-Darwinists had done), but also *including psychological properties*.

Brentano's concerted attempt to use science against Darwinian natural selection to make the case for the existence of God must have seemed to the young Freud, with his materialist orientation, to be an invasion of science by religion.

For a while, caught up in the magic of Brentano's seductive philosophical theism, Freud, according to his own testimony, actually abandoned materialism. He did so, however, without ever embracing theism. It was not long before he threw off Brentano's influence, and returned to materialism. Brentano's lectures *On the Existence of God*, for all their cleverness, went no further than to argue the case for some kind of intelligent design. Freud, no doubt with Brentano in mind, was later to dismiss all such abstract, philosophical conceptions of religion as mere "rearguard actions," which avoided the real issue: the projection of God in the form of an exalted father.[17]

The History of an Illusion

Freud's first major work on religion was *Totem and Taboo*, published in 1913, and his last written work was *Moses and Monotheism*, which appeared shortly before his death in 1939. Both of these books were concerned with what Freud referred to as "the historical truth" of religion. The essence of Freud's view that the intimate relation between obsessional neuroses and religious practices, which he had detected as early as 1907, had an actual historical point of origin was best expressed in *Moses and Monotheism*:

> Early trauma—defence—latency—outbreak of the neurosis—partial return of the repressed material: this was the formula we drew up for the development of a neurosis. Now I will invite the reader to take a step forward and assume that in the history of the human species something happened similar to the events in the life of the individual. That is to say, mankind as a whole also passed through conflicts of a sexual-aggressive nature, which left permanent traces, but which were for the most part warded off and forgotten: later, after a long period of latency, they came to life again and created phenomena similar in structure and tendency to neurotic symptoms.
>
> I have, I believe, divined these processes and wish to show that their consequences, which bear a strong resemblance to neurotic symptoms, are the phenomena of religion.[18]

To make sense of this we have to recognize that Freud was an evolutionary thinker, strongly influenced by Darwin, who nonetheless held on to two non-Darwinian evolutionary views prominent among scientists in his day that were to be discredited in later evolutionary theory: the recapitulation theory and Lamarckian evolution. The recapitulation theory, most directly associated with Darwin's leading German follower, Ernst Haeckel, was the view that ontogeny recapitulates phylogeny. That is, each individual of the species in its development passes through (recapitulates) in telescoped fashion the main stages that the entire species over historical time had previously passed through. As Freud put it in his *Introductory Lectures on Psychoanalysis* (1916–17), "Each individual somehow recapitulates in an abbreviated form the entire development of the human race."[19]

Lamarckianism was the notion that evolution proceeded through the inheritance of acquired characteristics. Frank Sulloway, in his *Freud, Biologist of the Mind*, described Freud as a "zealous psycho-Lamarckian" throughout his career. During the First World War he even began writing a work in collaboration with Sándor Ferenczi that was intended as a major contribution to psycho-Lamarckianism and its relation to psychoanalysis (a project that was soon abandoned, however).

Freud's adherence to Lamarckianism was not altogether at odds with the science of his day. In Freud's scientific generation, most biologists, as we have noted, were Lamarckian to a degree. Even Darwin introduced elements of Lamarckianism into later editions of *Origin of Species*. But in embracing the notion that the internal *needs* of organisms and their attempts at satisfying them were the principal bases of evolution, with the drive to fulfill such needs resulting in new characteristics encoded for future generations, Freud adopted the aspect of Lamarckianism most criticized by Darwin. Darwin called this the "Lamarck nonsense of a 'tendency to progression,' 'adaptations from the slow willing of animals,' &c.!" It allowed evolution to proceed much more straightforwardly and rapidly than in Darwinian natural selection.[20]

Moreover, Freud held on to Lamarckian evolutionary conceptions after their scientific popularity was in sharp decline. Thus in

Moses and Monotheism he commented on science's growing disfa-
vor toward Lamarckian conceptions by referring to "the present
attitude of biological science, which rejects the idea of acquired
qualities being transmitted to descendants. I admit, in all modesty,
that in spite of this I cannot picture biological development proceeding without taking this factor into account." [21]

Both recapitulation and Lamarckianism were integral to
Freud's scientific argument in *Totem and Taboo* and *Moses and
Monotheism. Totem and Taboo* was subtitled *Resemblances Between
the Psychic Lives of Savages and Neurotics.* It had its basis in
insights from Darwin and early anthropological studies. In *The
Descent of Man* Darwin argued that humans were originally organized in small groups around a dominant male. As he stated:

> If we look far enough back in the stream of time, it is extremely improb-
> able that primeval men and women lived promiscuously together.
> Judging from the social habits of man as he now exists, and from most
> savages being polygamists, the most probable view is that primeval man
> originally lived in small communities, each with as many wives as he
> could support and obtain, whom he would have jealously guarded
> against all other men. Or he may have lived with several wives by himself,
> like the Gorilla; for all the natives "agree that but one adult male is seen
> in a band; when the young male grows up, a contest takes place for mas-
> tery, the strongest, killing and driving out the others, establishes himself
> as the head of the community."[22]

In addition to Darwin, Freud relied on the work of a number of
British anthropologists/ethnologists, particularly W. Robertson
Smith and J. G. Frazer. Smith published his *Lectures on the
Religion of the Semites* in 1889 and was one of the early analysts of
totemism (totem means kinship). In the *Religion of the Semites* he
focused on the so-called totem feast with its sacrifice and eating of
the totem animal and the existence of two primary taboos associated with totemism: the taboo against murder and the taboo against
incest, i.e., the institution of exogamy whereby totem members

were only allowed sexual relations outside the totem group. This analysis was then carried forward by Frazer, author of *The Golden Bough: A Study in Comparative Religion* (1890) and *Totemism and Exogamy* (1910).[23]

Based on the clues offered by Darwin, Smith, Frazer, and others, together with Lamarckian evolutionary conceptions and Haeckel's recapitulation theory, Freud theorized a complex phyletic past in which the Oedipus Complex in modern children and adult neurotics could be traced back over the millennia to primitive cultures.

In the final chapter of *Totem and Taboo* Freud proposed that the historic event that gave rise to civilization was the murder of the primal father by his sons. Darwin's conception of primeval human societies as small groups ruled over by a dominant male gave rise to what Freud called the "Darwinian primal horde," according to which the lead male ruled over the females of the group who were his property and consorts. Sexual access to the women in the horde was therefore denied to the other males, most of whom were the sons of the dominant male. Those sons that posed a threat to the father's exclusive sexual rule were driven out, castrated, or killed.[24] If a son rebelled and overthrew the father, he simply ended up reproducing the same system by assuming the role of patriarch himself.

The historical deed that changed all of this, resulting in the first step toward civilization, according to Freud, was the formation of a band of brothers who, acting together, overthrew and killed the father, and then, "according to the custom of those times—all partook of his body." The patricide gave way almost immediately to feelings of guilt. The brothers therefore established a covenant whereby they renounced sexual access to the females in the horde—turning away from the goal that had motivated their rebellion. The new form of social organization that arose out of the brother band was expressly designed to prevent a return to the primal horde by establishing taboos with respect to murder and incest.

For Freud it was significant that these two taboos corresponded "with the two crimes of Oedipus, who slew his father and took his mother to wife, and also with the child's two primal wishes whose insufficient repression or whose re-awakening forms the nucleus perhaps for all neuroses." Freud's story of primal patricide thus appears to explain at one and the same time the psychogenesis of a child's primal wishes, adult neuroses, and religious practices. Freud introduced the Oedipus Complex into psychoanalysis, using that term, in 1910. It could now be traced to a historic deed, the murder of the primal father, upon which civilization was erected.[25]

Totemism, Freud believed, was a universal stage of human development that grew out of these traumatic events. The totem feast, in which the totemic animal, the object of worship, was killed and eaten, was, in his conception, a symbolic reenactment of the murder and the devouring of the primal father. Totemism, with its hard and fast taboos against murder and incest, thus constituted the first religion, arising out of a primeval patricide.

Totemism is followed in Freud's theory by a period of religious development in which the totem animal, symbolizing the father, is given human form, in the deities and heroes of antiquity, leading eventually to the paternal monotheistic god of the Judaeo-Christian tradition in particular, constituting the return of the repressed father.

In his final work, *Moses and Monotheism*, Freud depicted the murder of Moses by his followers as playing a similar role in the foundation of Judaeo-Christian religion.[26] Freud was convinced by a monograph by biblical scholar and archaeologist Ernst Sellin, who claimed to have found evidence in the book of the Prophet Hosea that Moses had been murdered by the people he had liberated and led from Egypt.

Inspired by Sellin's account, Freud put forward the conjecture that Moses was an Egyptian, who had introduced the Israelites to monotheism derived from the Egyptian worship of Aton (the sun god), and then led them out of Egypt. His adopted people, however, ended up murdering him following the Exodus and abandoned his religion. Yahweh remained the god of the Jewish people for as much as eight centuries (a period of cultural latency). And then there was a "return of the repressed," as a new prophet arose, who, assuming the name of Moses from the earlier reformer, reestablished the Mosaic religion.[27] In this account, we find once again, as in *Totem and Taboo*, that a patriarchal figure (and in this case founder of a religion) is murdered, and the remorse after a period of latency results in a return of the repressed, leading to a new (monotheistic) religion.

One of the characteristics of the Mosaic religion, according to Freud, was the prohibition against "making an image of God, which means the compulsion to worship an invisible God." Since this "signified subordinating sense perception to an abstract idea; it was a triumph of spirituality over the senses." God became "dematerialized" in the Jewish religion, increasing the importance of "their written records."[28]

Christianity, according to *Moses and Monotheism*, arose when a Jew, Saul of Tarsus (Saint Paul), intuitively grasped that the "original sin" of the murder of Moses, and before that the murder of the primal father, lay beyond neuroses evident in civilization. He therefore founded a dynamic cult based on the sacrificial death of Jesus, who had embraced the guilt of the world. "A Son of God, innocent himself, had sacrificed himself, and had thereby taken over the guilt of the world. It had to be a Son, for the sin had been the murder of the Father." Christ's sacrifice came to be reenacted in the form of holy communion (a transformation of the totemic feast). The death of the father behind the notion of God is therefore atoned for by the death of the son, generating a religion of the Son. Christ is thus both a substitute for and successor to Moses. Christ was at one and the same time the "resurrected Moses" and the "returned primeval father." Representing a step forward in the "return of the repressed," Christianity, according to Freud, displaced Judaism, reducing the latter to a mere "fossil."[29]

Although religion was always, for Freud, an illusion, i.e., a system of belief founded on wish-fulfillment, his history of this illusion in *Totem and Taboo* and *Moses and Monotheism* led him to conclude that behind it lay a concrete "historical truth." Thus, as he wrote in the latter work:

> I too should credit the believer's solution [i.e., "the idea of an Only God"] with containing the truth; it is not, however, the material truth, but a historical truth. I would claim the right to correct a certain distortion which this truth underwent on its re-emergence. That is to say, I do not believe that the one supreme great God "exists" today, but I believe that in primeval times there was one person who must needs appear gigantic and who, raised to the status of a deity, returned to the memory of men.[30]

Neither the "historical truths" (the primal horde, the totemic feast) nor the biological truths (recapitulation and Lamarckianism) that underlay Freud's theory of the psychogenesis of religion have withstood the tests of time and science. Both Darwin's notion of the primeval horde and Smith's totemic feast have been questioned, and their universality rejected by later anthropology. As Peter Gay notes:

> Cultural anthropologists demonstrated that while some totemic tribes practice the ritual of the sacrificial totem meal, most do not; what Robertson Smith had thought the essence of totemism turned out to be an exception. Again, the conjectures of Darwin and others about the prehistoric horde governed autocratically by a polygamous and monopolistic male did not stand up well to further research, especially the kind of research among the higher primates that had not been available when Freud wrote *Totem and Taboo*.[31]

Likewise, Sellin's argument on the murder of Moses was not well documented (as Freud conceded) and was later abandoned by Sellin himself. It no longer has any influence.[32]

Worse still, Freud's whole notion of the psychogenesis of religion developing as a result of an unconscious cultural memory that was carried forward in subsequent generations, and that took on the same form (passing through the same stages) as now evident in adult neuroses, depended on the twofold notions of recapitulation and Lamarckian evolution. More concretely, the proposition that the childhood Oedipus Complex could be traced back to the over-throw of the father clan by the brother band due to the desire for sexual access to the females of the horde depended on this same twofold recapitulation/Lamarckianism. Yet, both recapitulation and Lamarckianism were rejected in twentieth-century biology, removing the scientific foundations of Freud's theory of the origin of religion. As biologist Stephen Jay Gould concluded his assess-ment of what he called Freud's "phylogenetic fantasy," "Freud's theory ranks as wild speculation, based on false biology and root-ed in no direct data at all about phylogenetic history."[33] Its influ-ence today has therefore been reduced to that of cultural symbol and metaphor.[34]

The Future of an Illusion

If Freud's critique of religion had been simply restricted to his attempt to generate a psychogenesis of religion it might be possible to dismiss his whole treatment of religion as a fascinating relic in the history of science. Yet, Freud also engaged in the critique of religion in a contemporary context. His best-known work in this respect is *The Future of an Illusion*, which appeared in 1927. Here Freud referred back to his special theory of the origin of religion in *Totem and Taboo*. But the real thrust of *The Future of an Illusion* was a wider philosophical assessment of religion as an illusion and its con-flict with science, building on earlier materialist critiques of religion.

The Future of an Illusion was concerned with what Freud called "religious ideas in the widest sense—in other words . . . in its illu-sions." Religion, he argued, is a form of wish-fulfillment, and beliefs erected on this basis are necessarily illusions, but not necessarily delusions. Religious teachings are thus beliefs that do not rest on experience or evidence. They do not, as in science, await confirmation, and are "insusceptible of proof," giving rise to "intellectual atrophy." Sacred religious writings, he argued, pointing to the tradition of critique going back to figures like Reimarus, Strauss, and Bauer, have been shown to be full of contradictions. Hence, the Christian father Tertullian, in declaring faith in God to be rooted in the acceptance of the absurd, captured the essence of faith.

"An illusion," Freud emphasized, "is not the same thing as an error, nor is it necessarily an error." Columbus's journey, he notes, was based on an illusion, but produced important results. "What's characteristic of illusions is that they are derived from human wishes. . . . For instance, a middle-class girl may have the illusion that a prince will come and marry her. This is possible; and a few such cases have occurred. That the Messiah will come and found a golden age is much less likely. Whether one classifies this belief as an illusion or as something analogous to a delusion will depend on one's personal attitude."

The irony of religion, and particularly of a monotheistic religion such as Christianity that promises heavenly salvation, is that it presents the world and the hereafter (required by the reality of death) *just as we would wish it to be.* "We shall tell ourselves that it would be very nice if there were a God who created the world and was a benevolent Providence, and if there were a moral order in the universe and an after-life; but it is a very striking fact that all of this is exactly as we are bound to wish it to be."

Indeed, to understand religion, Freud believed, one had to perceive the psychological functions it fulfilled. He depicted God as performing three functions for society: (1) exorcising cosmic terrors; (2) reconciling individuals to the cruelties of fate; and (3) compensating for the privations necessary for the workings of society. Yet, to demand that God fulfill these functions was to live in an infantile state. To give way to the illusions of God was to fall prey to a state of mere wish-fulfillment rather than to rely in a more adult fashion on cold science.

Freud conceded that religion had "clearly performed great services for human civilization," particularly in the moral sphere. But in promoting morality it also promoted immorality. In promising the benevolence of God, it also spread the fear of God. "Where questions of religion are concerned," he wrote, "people are guilty of every sort of dishonesty and intellectual misdemeanour." Like Marx, Freud threw scorn on what he called Tertullian's "desperate" attempt "to evade the problem" in his: "I believe because it is absurd."[35]

Peter Gay called Freud "the last philosophe"—a direct heir to the Enlightenment critique of religion. Science, for Freud, required the direct challenge of the common person's religion (promoted by established religious hierarchies), based on the exalted patriarchal God. The Scopes "monkey trial" in Tennessee, according to Freud, had shown that Americans alone were "consistent" in recognizing the depth of the contradiction between science and religion. Freud was particularly concerned by what he referred to as growing instances of the "invasion by religion of the sphere of scientific thought."[36] There can be no doubt that he was alarmed by the implications for science of the Scopes "monkey trial." But it is equally evident that Freud, in referring to the "invasion by religion," was reflecting back on Brentano's stupendous efforts to construct a philosophical theism that used intelligent design arguments to counter the effect of Darwinian evolutionism and materialism.

In this war between religion and science there was no room for compromise. Religion sought to keep the faithful in an "infantile" condition. Yet, individuals could not remain children forever. Rather, they must "concentrate their energies into life on earth." Freud was thus fond of quoting the German poet (and friend of Karl Marx) Heinrich Heine: "We leave Heaven to the angels and the sparrows."[37]

Freud's general critique of contemporary religion was most powerfully and succinctly expressed in his *New Introductory*

Lectures on Psychoanalysis of 1933. Here he argued that the *Weltanschauung* of religion was being replaced by the *Weltanschauung* of science. Religion was more "grandiose" than science in that the former "leaves no question unanswered." It fulfilled "three functions" for human beings (related to the three functions that Freud had already given to the concept of God in *The Future of an Illusion*). "It gives them information about the origin and coming into existence of the universe, it assures them of its protection and of ultimate happiness in the ups and downs of life and it directs their thoughts and actions by precepts which it lays down with its whole authority." Due to the grandiose way in which it fulfills these three functions "religion alone is to be taken seriously as an enemy" of science, while science is hard-pressed to fulfill the same needs. But religion's great strength and also its weakness is that it is "insusceptible of proof." The scientific *Weltanschauung*, in contrast, is a way of employing the intellect that by its nature awaits, indeed demands, confirmation, and thus progresses in stages. It tentatively—but with even firmer logic and intolerance—establishes a "dictatorship in the mental life of man."[38]

Nevertheless, the struggle between religion and science, Freud argued, persists, since the supporters of religion claim that there is a realm of supreme knowledge (divine intelligent design) that mere science can never attain. As Freud summarized these attacks on science:

> The supporters of the religious *Weltanschauung* act upon the ancient dictum: the best defense is attack. "What" they ask, "is this science which presumes to disparage our religion?...Can it tell us how the universe came about and what fate lies before it? Can it even draw us a coherent picture of the universe, or show us where we are to look for the unexplained phenomena of life or how the forces of the mind are able to act on inert matter?...It gives us fragments of alleged discovery, which it cannot bring into harmony with one another; it collects observations of uniformities in the course of events which it dignifies with the name of laws

and submits to its risky interpretations. And consider the small degree of certainty which it attaches to its findings! Everything it teaches is only provisionally true: what is praised to-day as the highest wisdom will be rejected to-morrow and replaced by something else, though once more only tentatively. The latest error is then described as the truth. And for this truth we are to sacrifice our highest good!"[39]

According to Freud, the invasion of science by religion in the name of creationism/intelligent design, symbolized by the Scopes "monkey trial," was nothing but a counterattack of religion against the inroads of science on what it considered its domain. Religion claimed on ostensibly empirical, rational grounds that mere science was forever incapable of grasping the full complexity or the meaningfulness of the world. The explanation for this offered by the religious *Weltanschauung* was that the divine was beyond the reach of the scientific method. Indeed, the whole goal of the religious invasion of science, Freud implied, was to establish boundaries to the pretensions of the latter. Religion mocked science for its tentativeness and its unknowns, claiming that these were indications of God's unfathomable purpose.

Freud's response was that science was still young, and its necessary "dictatorship in the mental life of man" would progress, uncovering further secrets of the material world by materialist means. The tentative and uneven, but inexorable, progress of science was only just beginning. As evidence of scientific materialism's youth, Freud pointed out that he himself "was already alive when Darwin published his book on the origin of species," and the time that had transpired since ancient Greek materialism was "only a small fraction of the length of time which anthropologists require for the evolution of man from an ape-like ancestral form, and which certainly comprises more than a hundred thousand years."[40]

Freud was not above borrowing from the language of religion in his critique of religion. Thus he declared that his own deity was the God of "Logos" or Reason. Religion, insofar as it mattered, that is, inasmuch as it took the form of the worship of an exalted patriarchal god, was to be regarded "not as a permanent acquisition" of civilization. The bondage of religion would be supplanted by the liberation offered by science. "From that bondage," Freud grandly declared to his students, "I am, we are, free. . . . No, our science is no illusion. But an illusion it would be to suppose that what science cannot give us we can get elsewhere."[41]

8. In Defense of Natural Science

The same kinds of attacks on science by religion that Freud described can be seen today emanating from those who attempt to use rational, empirical arguments to demonstrate: the limits of reason and science; the reality of design in the universe; and the necessity of a religious *Weltanschauung*. Insofar as materialist-scientific understandings are rooted in notions of contingency, evolution, and emergence—as materialists since Epicurus have argued—intelligent design proponents insist such views are based on "mere chance," and are incapable of comprehending an irreducibly complex world arising from a higher intelligence.

Marx, following Hegel, ironically observed that though traditional theology saw the divine world in the form of innumerable supernatural accidents or miracles emanating from God, natural theology rejected accident in favor of a pervasive intelligence, which was then taken as evidence of God.[1] In line with this, proponents of natural theology and intelligent design throughout history have decried the role of chance and contingency in materialist thought, arguing that it is insufficient, even when combined with a process of natural selection, to explain the organization of life.[2]

Darwin's theory of evolution overthrew the devout contemplation and divine wonder of special creation, offering in its place a means to understand the material world in its unfolding grandeur by studying its historical patterns and natural mechanisms. This materialist revolution challenged theological claims that the spirit of God pervaded the universe, that morals and values were innate and sacrosanct, and that faith and scripture rather than reason and empirical evidence are the basis for establishing truth. Scientific knowledge retains a radical dimension stemming from this, opening up the world to human inquiry. In contrast, intelligent design proponents are in an inherently reactionary position, offering supernatural "design" as the ultimate explanation for the world, placing faith before science.[3]

The Logic of Design

The logic of the intelligent design argument is straightforward. Instead of seeking to provide a wealth of answers of scientific interest, it is concerned with promoting a single *design inference*: "God did it." The design hypothesis, as biochemist and senior fellow of the Discovery Institute's Center for Science and Culture, Michael Behe, explains, tells one nothing about actual natural mechanisms. "If the designer was in fact God," he writes, "then there is good reason to suppose that the mechanism of design will forever remain beyond us." Further, if God is the designer, Behe observes, a distant miracle may be presumed; but the intelligent design argument can say nothing about the miracle itself. The only positive result of the intelligent design argument is therefore the inference of intelligent design itself, behind which lies an Intelligent Designer. It can say nothing about either the design process or the designer other than that they exist.

The whole point, moreover, is to single out the agency of a designer, as distinguished from mere natural laws, viewed as designed. "As soon as design is located in natural laws," Behe's colleague William Dembski acknowledges, "design becomes an empty metaphor. . . . a superadded principle devoid of empirical content." Only by pointing to a designer (and not simply laws that are designed) is it possible to build a "bridge between science and theology."[4]

Specifically, intelligent design proponents argue that many features of the natural world, particularly biological structures, are too complex to arise from naturalistic causes. "Intelligent design," in the words of Stephen Meyer, program director of the Center for Science and Culture "holds that there are tell-tale features of living systems and the universe that are best explained by an intelligent cause." He notes that DNA is like a software program or "an advanced form of nanotechnology" and must have had a programmer/designer. Given that it holds information, something with intelligence must have played "a role in the origin of DNA." In fact, "we know from experience that systems possessing these features invariably arise from intelligent causes." Meyer is thus content to assert as a scientific argument that "intelligent design," *which itself is beyond explanation*, "best explains the origin of molecular machines within cells." He concludes: "living organisms look designed because they really were designed" by a designer.[5]

In this way, intelligent design supporters attempt to present their thesis as an answer to scientific questions. Yet, the key proposition of intelligent design lies outside the realm of scientific inquiry, since it cannot be adjudicated with empirical evidence. As David Hume stated in his critique of natural theology in his *Enquiry Concerning Human Understanding* (as noted in chapter 4 above), to argue in this way is to "embrace a principle, which is both uncertain and useless. It is uncertain; because the subject lies entirely beyond the reach of human experience. It is useless; because our knowledge of this cause being derived entirely from the course of nature, we can never, according to the rules of just reasoning, return back from the cause with any new inference."[6] Indeed, not only are intelligent design proponents unable to provide convincing scientific evidence that a given phenomenon must have been designed, the mere assertion, on their part, creates a "boundary condition" to science and therefore inhibits its further progress.[7]

Like Meyer, Behe claims that evidence of intelligent design is to be found in the apparent rational design of natural systems—from the structure of DNA and the machinery of cells to the integration of organisms and the harmony of ecosystems. Certain natural systems can be characterized as "irreducibly complex." By this, Behe means that biological systems are composed of "interacting parts that contribute to the basic function" and operation of the system as a whole. He contends that insofar as a given system is irreducibly complex such organization cannot emerge directly from "successive modifications of a precursor system," because if any of the parts are missing from such a system it would cease to function. Therefore, he infers, design is necessary to create such functional, irreducibly complex biological systems.[8]

Intelligent design proponents like Behe and Meyer marvel at the wondrous information DNA contains, noting how it operates like a computer, processing information, maintaining the needs and operations of living systems. They contend that too much information is contained in DNA for it to have developed by blind forces, or chance, just as they insist that parts of organisms, such as the bacterial flagellum, the human eye, and the human brain, are too complex to have been designed by the piecemeal tinkering of natural selection.

Referring to cells and DNA this view takes the same general form as the classic argument from design, assuming the complexity of life is beyond materialist explanation, while at the same time pretending to engage in serious scientific investigation/inference. Although the intelligent design textbook *Of Pandas and People* presents some new-fangled creationist ideas in which evolution is denied altogether and "change is limited to variation within existing groups of plants and animals," today's newfangled intelligent design creationism is typically more subtle.[9] They do not *necessarily* deny evolution entirely, or insist on the rigid separation of species and the notion that no intermediaries are possible. Instead, they concentrate on finding *just a few* irrefutable examples of "irreducible complexity," the best-known example being the bacterial flagella. Intelligent design proponents thus refuse to be pinned down, giving way to evolutionary theory on innumerable occasions, but looking for some critical complexity gap that evolutionary theory cannot account for, operating under the assumption that this is all that is needed to make the "design inference" plausible. As John G. West, associate director of the Center for Science and Culture, has written:

> Strictly speaking intelligent design is not "anti-evolution." It does not challenge the idea that living things "change over time," nor does it deny that Darwin's mechanism of natural selection can produce changes in living things. It is not incompatible with the idea that all living things arose from a universal common ancestor, although scientists who support intelligent design differ on whether the scientific evidence substantiates such a conclusion. Intelligent design does oppose the central claim of Darwinian evolution that all of the highly specified complexity in nature can be accounted for through an undirected process such as natural selection acting on random variations.[10]

According to this view, the intelligent design argument rests simply on countering the Darwinian claim that *all* instances of complexity in the natural world can be explained in evolutionary terms. In this way materialist science is placed on the defensive and forced into the position of perpetually offering "stopgap" arguments for all gaps in current knowledge.[11] Any failure of science to close absolutely every gap in evidence is treated as an absolute victory for design. Even on these terms, however, the theory has been an abysmal failure. Its focus, particularly in the work of Behe, on "irreducibly complex systems," such that the removal of one part/function would cause the system to fail, has again and again proven to be empirically and logically inadequate, since the real point, in terms of the critique of natural selection, is whether such "irreducibly" complex systems could have originated through natural processes. In each case raised (such as the bacterial flagella) evolutionary paths have been demonstrated in the scientific literature.[12] In his self-criticism Behe has admitted to this failure, calling it the "asymmetry" of his argument, which fails to make a case for its primary inference: that evolution could not generate instances of "irreducible" complexity.[13]

In his latest book, *The Edge of Evolution*, Behe has shifted his argument to focus more on the random nature of mutation that provides the raw material upon which natural selection operates.[14] He argues that random mutation simply does not provide sufficient variation to allow for dramatic evolutionary change, although it may account for minor modifications of organisms. Instead, it is an unspecified designer, acting as a genetic engineer, who orchestrates the processes giving rise to new species. In this unsupported contention about mutation, Behe contradicts the scientific consensus based on the research of a long line of mathematical geneticists (some of whom, such as J. B. S. Haldane and Richard Lewontin, were influenced by the historical materialist tradition), which has clearly established that the rate of naturally occurring mutation exceeds that which is necessary for natural selection to produce the full range of organisms we observe.

Irreducible Complexity and Design

Intelligent design proponents have focused much attention on bacterial flagella—"propellers" on some bacteria that are used for mobility. It is claimed that these propellers are too complex to have been produced by evolutionary processes, given that they involve separate parts that need to be assembled to fulfill a specific collaborative function. This argument in many ways is logically similar to the one that has long been raised against the evolution of the human eye, which has been shown to be invalid. Behe touted bacterial flagella at the 2005 Dover trial as an example of irreducible complexity that could not be accounted for except through the actions of a designer. This has been thoroughly refuted in the scientific literature and was shown to be false during the trial in the testimony of several scientists, including Kenneth Miller, a professor of biology at Brown University. The design argument, which insists upon the creation of separate flagellins (separate lineages of bacteria with flagella), implies "there were thousands or even millions of individual creation events, which strains *Occam's razor* to the breaking point." In contrast, evidence indicates that natural selection has produced the variety of distinct flagellins—in other words, "the highly diverse contemporary flagellar systems have evolved from a common ancestor."[15]

Unwilling to acknowledge this contrary evidence, Behe, particularly in *The Edge of Evolution*, asserts that an Intelligent Designer must have intervened in the evolutionary process to produce the myriad forms we see, which cannot be attributed to chance or random mutation.[16] Yet, one of the key points he fails properly to appreciate, although he is undoubtedly aware of it, is that natural selection, not random mutation, is the creative force in Darwinian evolution. Although mutation in many ways may be regarded as random, natural selection is anything but random; it serves systematically to preserve genetic mutations that enhance the reproductive success of their host organisms.

William Dembski, who next to Behe is the most important proponent of intelligent design's science argument, attempts to argue that the improbability of certain chance results makes the "design inference" more likely.[17] These contentions, however, are based on the confusion of pure chance (like the rolling of dice) with contingency based on evolutionary pathways and interactions. By presenting evolutionary theory in a mechanical, reductionist form, and alleging that its explanations of organic evolution depend on the action of pure chance, intelligent design proponents misconceive science and make their arguments for design seem more plausible. Thus Dembski repeatedly suggests, though he should know better, that there is nothing logically and materially in between pure chance, e.g., the rolling of dice, and design. "What laws cannot do," he writes, "is produce contingency. . . . If not by means of laws, how then does contingency . . . arise? Two and only two answers are possible here. Either the contingency is a blind purposeless contingency, which is chance; or it is a guided, purposeful contingency, which is intelligent causation." What Dembski's "two and only two answers" leaves out, of course, is that contingency, distinct from pure chance, operates along historically and structurally conditioned pathways, i.e., the reality of material evolution, the hallmark of the modern scientific perspective.[18]

Structure and Dialectics: Necessity and Contingency

Marx's dialectics of nature and society continues to serve as a powerful base for a critique of intelligent design. It includes a commitment to a materialist conception of natural and social history—and thus to the interaction of necessity and contingency. From this tradition, which influenced noted scientists such as J. B. S. Haldane, J. D. Bernal, Joseph Needham, and many others, in the 1930s and '40s, and Stephen Jay Gould, Richard Lewontin, and Richard Levins, among numerous others in more recent years, we can gain insights into the dynamic development and interaction of society and nature.[19] Evolutionary theorists have developed explanations of how evolution can produce forms that *superficially appear* irreducibly complex through the operation of purely materialist forces. The structuralist tradition in biology, although it has often been invoked by intelligent design proponents in support of their arguments, provides concepts that, when infused with dialectical insights, compellingly explain how many complex features are evolved.

The argument about bacterial flagella—raised at the Dover trial—is part of the more general challenge to evolution that has been raised by creationists since the publication of the *Origin of Species*. It is the problem of incipient stages—i.e., how integrated complex structures evolve when less complex versions of them appear not to serve the same function to some degree. Since natural selection is a "blind" process, with no foresight, it does not construct features of organisms with a goal in mind. Rather, natural selection preserves chance mutations that enhance fitness. Thus, based on a superficial assessment, it seems hard to explain how an integrated complex structure can evolve, unless its nascent steps are in themselves adaptive.

Darwin and subsequent Darwinians have provided compelling explanations for how a variety of complex features evolved. The basic argument is well illustrated in Richard Dawkins's explanation of the vertebrate eye: nascent features can indeed provide adaptive advantage similar to the more complex structures into which they may later evolve, although to a lesser degree. An "eye" that does not allow one to focus may still be useful in detecting changes in light and movement. Dawkins writes, "Vision that is 5 per cent as good as yours or mine is very much worth having in comparison with no vision at all. So is 1 per cent vision better than total blindness. And 6 per cent is better than 5, 7 per cent better than 6, and so on up the gradual, continuous series."[20] Cases where nascent and intermediate stages provide adaptive advantage similar to the later evolved feature are fairly easily understood within an evolutionary framework, since it is easy to see how a complex structure can be built in steps by natural selection when each step improves fitness. However, when the adaptive advantage of a nascent stage is unclear, more subtle explanations are required.

The more complex line of argument, which addresses these more difficult features, invokes "functional shifts," recognizing that over the course of evolutionary history the use to which a structure is put may change. Thus, nascent and intermediate stages of a structure may have been selected for because they served different purposes at different points in time in a lineage's phylogeny (evolutionary history). Although this is an old argument, going back to Darwin, it was conceptually enriched in the latter part of the twentieth century, particularly in the work of Stephen Jay Gould and his close colleagues, who drew extensively on the insights coming from the structuralist tradition in biology.

Gould points out that organisms are not mere putty to be sculpted over the course of their phylogeny by external environmental forces, but, rather, their structural integrity constrains and channels the variation on which natural selection operates.[21] In this, Gould challenges the notion that phenotypic variation is isotropic, equally likely in all directions. Although mutation produces genetic variation that is random relative to selective advantage, this does not mean that phenotypic variation is not more likely in some directions than in others. Gould notes that the structural nature of the development of an organism throughout its life course (ontogeny) limits the types of phenotypic variation that are possible, because changes at one stage of the developmental process have consequences for later stages. Therefore, many characteristics of an organism cannot simply be modified without having substantial ripple effects throughout the whole organism. The inherited patterns of development, therefore, do not readily allow for all types of modification. Hence, the evolutionary process is a dialectical interaction between the internal (inherited structural constraints) and the external (environmental selection pressure), just as the ontogeny (development over the life course) of individual organisms is a dialectical interaction between their genes and the environment. Such an understanding helps restore the organism as a concept in biology—"an integrated entity exerting constraint over its history" while being situated in a specific environmental context.[22]

The structural nature of development has consequences for patterns of change. To illustrate this point, Gould makes use of a metaphor: Galton's polyhedron.[23] As he frequently does, Gould draws upon the arguments of various historical figures involved in evolutionary debate to build his own. Francis Galton, who was Darwin's cousin (Erasmus Darwin was grandfather to both) and is regarded as the father of eugenics, was deeply impressed by his cousin's work on evolution, but he disagreed with Darwin's assumptions about the nature of variation. He developed a metaphor to challenge aspects of Darwin's conception of natural selection and the nature of change. Adopting Galton's conceptual insight, Gould explains that in the idealized Darwinian formulation species are metaphorical spheres that roll freely on any phylogenic course through morphospace that the external world pushes them along—i.e., they do not have a structural integrity that offers resistance to pressure from the external environment, and thus they move readily wherever environmental forces direct them via natural selection. Alternatively, in the metaphor of Galton's polyhedron, species are polyhedrons, multi-sided solid objects that have flat faces (such as dice), whose structure prevents them from rolling freely when only slightly perturbed and limits the paths they can follow after receiving a sufficient push from the external world. In other words, "change cannot occur in all directions, or with any increment." Polyhedrons can switch the facet on which they rest, but they cannot simply rest in any given position (e.g., they can rest on a face but not on a corner). In contrast with a sphere, which may roll smoothly with a light tap, the polyhedron will resist minor perturbations, but, given sufficient force, will switch facets abruptly—potentially generating changes "that reverberate throughout the system." Thus species cannot perfectly track changing environments because of the structural interconnections they develop over the course of their phylogeny that limit and, potentially, direct the type of change that is possible.

The key insight of Darwin was that structural constraint, rather than being God-given and immutable, is the product of evolutionary history. Gould emphasizes the importance of both recognizing the reality of structural constraint and also that structures have historical origins.[24] This perspective helps unite the insights from both sides of the age-old debate between functionalist biologists, such as Darwin, Lamarck, and Georges Cuvier, and formalist (structuralist) biologists, such as Étienne Geoffroy Saint-Hilaire, Richard Owen, and Johann Wolfgang von Goethe. Whereas the functionalists emphasized that features of organisms existed for

utilitarian reasons (e.g., they were adaptations to their environments), formalists emphasized the structural unity of type common
across similar organisms. Formalists, like modern-day creationists,
typically denied the possibility of evolution because they believed
that only superficial change was possible, not fundamental change
of underlying structures. Thus, intelligent design advocates
Benjamin Wiker and Jonathan Witt draw upon the arguments of
prominent eighteenth- and nineteenth-century formalists (such as
Owen and Saint-Hilaire) in making their argument for the impossibility of evolutionary change in the structural features of organisms.[25] However, these arguments were undermined by Darwin
and subsequent evolutionists, who recognized that structures had
evolved, although after their emergence they may indeed constrain
the evolutionary pathways available to organisms (as the metaphorical polyhedron comes to rest on a particular facet). Thus, as Gould
notes, Darwin fundamentally reoriented the functionalist-formalist
debate, by adding a new dimension to the functional (active adaptation) and formal (constraints of structure) dichotomy: history
(contingencies of phylogeny).[26] Intelligent design supporters have
obviously missed the innovation, and continue to expound views
that have long been superseded.

Based on his recognition of the importance of structure, Gould
explains how the structural nature of organisms provides one of the
keys to understanding the emergence of many complex features. In
a famous article, Gould and Lewontin introduced the concept of
spandrels, a term borrowed from architecture, as a metaphor for
structural elements that exist for nonfunctional reasons.[27] In architecture, a spandrel is a space that emerges as a consequence of the
meeting of two different structural elements of a building, such as
the triangular spaces that appear where an arch is set within a rectangular feature. The spandrel serves no function and is not created for a purpose—it is simply the side effect of combining different
structural elements. Applying this metaphor to biology, Gould and
Lewontin note that many features of organisms do not exist for a

direct functional reason, but rather emerge as a side consequence of the structured nature of development. One of the clearest examples of a spandrel can be found in snail shells. As Gould explains, "Snails that grow by coiling a tube around an axis must generate a cylindrical space, called an umbilicus, along the axis." Some species "use the open umbilicus as a brooding chamber to protect their eggs," but the substantial majority does not. "[U]mbilical brooders occupy only a few tips on distinct and late-arising twigs of the [snail] cladogram [evolutionary tree], not a central position near the root of the tree," an observation clearly indicating that the umbilicus did not originate as an adaptation for brooding.[28] The umbilicus, therefore, clearly appears to be a spandrel—a structural side effect of a growth process where a tube is coiled around an axis, not a feature directly crafted by natural selection for adaptive reasons.

Another important concept, introduced by Gould and Elisabeth S. Vrba, that helps us to understand how complex features evolve is that of *exaptation*.[29] Exaptation refers to the utilization of an existing feature for a novel functional purpose.

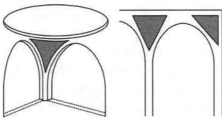

A pendentive, or three-dimensional spandrel (left), forms as a necessarily triangular space where a round dome meets two rounded arches at right angles. "Classical," two-dimensional spandrels (right), form as the necessarily triangular spaces between rounded arches and the rectangular frame of surrounding walls and ceilings.

The feature may either have originally evolved via natural selection to serve as an adaptation, in which case the co-optation for the new purpose represents a functional shift, or the feature may be a spandrel that did not evolve as an adaptation, but was rather simply a structural side effect. The umbilical brooding snails mentioned above represent a prime example of the latter, where the umbilicus did not originate for adaptive reasons, but was at some point after its origin utilized for brooding—a functional purpose. A prime example of the former is the shift in the primary function of feathers in the bird lineage from thermoregulatory to aerodynamic.[30] At any one time organisms have a vast "exaptive pool"—i.e., many features, both those that evolved under natural selection for the function they serve (adaptations) and those that are nonadaptive structural side effects (spandrels)—from which evolution may draw to craft new features as changing conditions dictate. Thus, structural forces can generate features, such as the umbilicus of the snail, that potentially provide well-formed starting points for the development of new adaptive traits. Functional shifts and the exaptation of spandrels thus can explain how the incipient stages of a feature may develop without serving the function to which they are later put.

Many features of the human mind illustrate well both spandrels and exaptation.[31] Undoubtedly, the mind has been shaped by natural selection to serve key functions. However, in the process, such a complex feature inevitably contains many spandrels that provide a vast exaptive pool. For example, just as a computer designed to perform mathematical calculations has the potential to be co-opted to produce video games, the human mind has the potential to perform many tasks that natural selection did not directly sculpt it for. We have the capacity to read and write, present mathematical proofs, and construct beautiful cathedrals, even though these capacities clearly played no role in the origin of the human mind and were not even actualized for the vast majority of human history. Complex features, like the mind, are inherently ripe with potential and have innumerable emergent properties. Features nearly always can be used in ways other than that for which natural selection produced them. When we recognize this structural point, and appreciate the theoretical importance of spandrels and exaptation, intelligent design arguments disintegrate and the material origins of life become apparent.

With an appreciation of the role of exaptation in evolutionary history, Mark J. Pallen and Nicholas J. Matzke clearly show how a feature as complex as the bacterial flagellum could evolve.[32] They detail how there are similar features with different functional purposes in other bacterial lineages. Thus, one need not assume that the flagellum was built piece by piece for its current purpose. Rather, it appears that key elements of it were exapted from forms that evolved for other reasons. This explains how a supposedly "irreducibly" complex system—which cannot serve its current purpose if it is missing even one of several components—can evolve in a piecemeal manner, rather than in "one fell swoop" as Behe has contended.[33] The truth is that bacterial flagella and countless other features in the natural world are the product of evolutionary pathways.

It is worth recognizing that there are a multitude of general sources of order in the biological world that can give natural selection a starting point from which to build complex features. In addition to theorists who focus on structures related to inherited patterns of development (historically developed structures), there is a tradition in biology that focuses on structural forces influencing organic form that stem from spatiotemporally invariant laws of physics, chemistry, and mathematics. The influence of such structural forces was extensively assessed by the polymath D'Arcy Wentworth Thompson in the early part of the twentieth century and remains a focus of active research among scholars today.[34] Thompson pointed to the ubiquity of features such as the logarithmic spiral (e.g., snail shells, ram's horns) and hexagonal forms (beehives and corallites) in nature, which appear to reflect general mathematical laws. Thompson argued that, just as crystals form in minerals following invariant physical and mathematical principles, many organized structures in the organic world are not the result of historical development.

Although it is generally agreed among modern biologists that Thompson greatly overextended his arguments, it is generally accepted that some forms, such as the spherical shape of individual cells, are not genetically encoded but rather formed through more general physical forces. The most prominent modern theorist in the tradition of Thompson is Stuart Kauffman, who has argued that much of the order we see in nature stems from spontaneous order that emerges in complex systems.[35] One of his most provocative suggestions is that a "cell type is an attractor of the genomic regulatory system."[36] In this sense, an *attractor* refers to a state toward which a dynamical system tends. In other words, Kauffman is suggesting that the diversity of cell types observed in organisms is not entirely a particularistic product of history, but rather the material outcome of the self-organizing principles of dynamical systems. This "order for free" is produced by an entirely naturalistic process, where the interaction of many forces leads to emergent structure without being designed, either by a supernatural force or natural selection. Such a position is in accord with the Epicurean tradition, which argues that organization itself creates the potential for the emergence of still higher forms of organization, "new modes of life, new functions or behaviors, impossible in less organized forms."[37] The spontaneous order that emerges in complex systems may provide initial structures in organic forms upon which natural selection builds.

The structuralist tradition in biology, upon which Gould and other dialectical biologists have drawn, provides many key insights that help explain the emergence of highly complex features, which may appear "irreducibly" complex to a naïve observer. It focuses on integrated organisms with specific histories and structures that interact with the dynamic, physical world. The combination of structuralism and dialectics is the key to understanding the evolutionary process and the development of the myriad wonders of the organic world. Natural laws and the contingencies of history explain complexity in nature, leaving no need for a designer and making God redundant in explaining the workings of the natural world.

9. Replaying the Tape of Life

The intelligent design movement is committed to interpreting the record of the history of life on Earth as showing signs of a designer. In the most obvious effort to find design in history, intelligent design supporters have tried to make much of the "sudden" appearance of new species in the fossil record, which creationists have long taken as evidence for the divine creation of species. More sophisticated versions of this doctrine will accept some level of evolution, but insist that God intervened at various points to direct evolution. The most sophisticated (and least falsifiable) of theistic interpretations of history do not invoke divine intervention after the initial creation, but argue that creation was rigged from the start to lead to the inevitable emergence of humans or some other intelligent being. These theistic perspectives are deeply committed to reading the patterns of history in a certain manner to fit their worldview, and thus the nature of history has long been a focus of intellectual battles over evolution. The materialist-dialectical tradition associated with Marxism has much to say on this topic, since from its inception it has been concerned with patterns of history and the tempo of change. In this chapter we discuss how material processes alone explain the patterns of history, including the apparently sudden appearance of species in the fossil record. History teaches us that the world was not predetermined but rather emerged through the interaction of chance and necessity.

Punctuated Equilibrium and Contingency

The discovery of "deep time" by geologists and of organic evolution by naturalists in the eighteenth and nineteenth centuries undermined the perspective favored by creationists, who were committed to a recent origin of Earth and saw order in nature that did not fundamentally change (e.g., species themselves did not change with time, although some may come and go at God's whims), although it was occasionally reshuffled by divine intervention (e.g., Noah's flood). Commitments to the fundamental stasis of the Creation with occasional changes brought about by divine acts gave way to the notion of slow, continuous change that was a key facet of the thinking of nineteenth-century scientists, reflected in Charles Lyell's uniformitarianism and Darwin's gradualism.[1]

Countering both the creationist view and the gradualist view, the Marxist conception of history is that change, governed by natural processes alone, is inevitable, but that it is not typically smooth and continuous, rather, often occurring very rapidly following periods of stasis (temporary periods, of indeterminate length, of counterbalancing opposing forces leading to relative stability). Naturally, neither the gradualist view of steady slow change nor the Marxian one of stability followed by occasional bursts of rapid change is absolutely correct; the complexity of human and natural history has ensured that both types of change occur. (It goes without saying that the rate of change is not binary, either necessarily rapid or gradual, but this dichotomy is heuristically useful.) Furthermore, the rate of change of any particular phenomenon is a factual question, and it cannot be determined without empirical evidence.

The unification of Darwinian and Marxian materialist views of historical change in the natural world is exemplified in Niles Eldredge and Stephen Jay Gould's argument that the evolutionary history of organisms is best characterized as "punctuated equilibrium"—long periods of stasis, punctuated with (geologically) brief periods of change.[2] This is based in part on a literal interpretation of the fossil record, which generally shows fossils of a species remaining quite similar over extended stretches of time, being suddenly (in the geological sense) replaced by a substantially different, although apparently related, type. Their argument is in no way a rejection of Darwinism in general, only a challenge to Darwin's strong preference for gradualism. Eldredge and Gould invoke no special mechanisms for change. Rather, they argue that speciation—the emergence of a new species—typically happens when a subset of a species becomes isolated. In a small isolated population, mutations can spread rapidly throughout the gene pool of the population, and the rate of change can be further accelerated if the population faces different selection pressures than the parent species. In large populations that are geographically widespread, although connected through breeding, mutations diffuse slowly, and any mutations that are favorable to organisms in one part of the range are not necessarily retained, since they may not be favorable to organisms in another part of the range. For these reasons, Eldredge and Gould proposed that widespread species will generally change little over most stretches of time, but may change rapidly around the point of speciation, when a subpopulation becomes isolated.

The theory of punctuated equilibrium inspired a vast amount of research in paleontology, genetics, and other fields related to evolutionary biology. This work established that Darwinian processes can and do indeed produce new species in geologically brief periods (e.g., tens of thousands of years), which will typically appear in the fossil record as instantaneous. The theory of punctuated equilibrium and the work it inspired undermined the creationist claim that the "sudden" appearance of new species in the fossil record contradicted evolution. Given that intelligent design proponents regularly argue that the gaps in the fossil record are evidence of intelligent design—suggesting divine intervention (a position known as "gap creationism")—they clearly recognize that the theory of punctuated equilibrium is directly antagonistic to their views. In fact, this interpretation of evolutionary theory is seen as the most threatening to them.[3]

It is worth noting that punctuated equilibrium, although conceptually independent, fits well with the structuralist views explained in the previous chapter. The metaphor of Galton's polyhedron illustrates how the structural integrity of forms, like the polyhedron, implies that there will be some degree of general stability, but when change comes it will happen fairly rapidly as processes cascade once set in motion (like the polyhedron changing facets). Chance and necessity interpenetrate and are involved in the process of macroevolution.

In addition to developing theories addressing the tempo of change in evolution, dialectical biologists also present a larger theoretical framework and more general materialist worldview that explain history as a contingent process governed by natural forces that is free of design and purpose. Gould best explains this materialist view of history in his renowned book, *Wonderful Life*.[4] He focuses on the fossils in the Burgess Shale, which are some of the earliest representatives of multi-cellular animal life from the middle Cambrian period, over 500 million years ago. He notes that the diversity of basic body plans present among these animals was at least as great as it is today, and that the species that left descendants that survived to the present day show no sign of being particularly dominant or otherwise "marked" for success in the Cambrian. For example, *Pikaia*, a two-inch-long wormlike creature that is the first known chordate, the phylum to which *Homo sapiens* and all other vertebrates belong, appears to have been of little significance in the Cambrian seas, being neither abundant nor otherwise remarkable. This is an important point, as Gould argues, because nothing about the history of life appears foreordained, least of all the rise of our own lineage.

Since natural selection adapts species to their local environment, it provides no particular direction or long-term trajectory to the evolution of species. Given that environments change in unpredictable ways, sometimes rapidly, the future history of life is unpredictable. Lineages that over the long haul of evolutionary history go extinct and those that survive are not typically distinguished from one another in a systematic way. Rather, historical contingency is capricious; and nature is indifferent. Consider, for example, the asteroid or comet that collided with Earth 65 million years ago at the end of the Cretaceous period, causing the extinction of dinosaurs. Dinosaurs had survived alongside mammals for nearly a hundred million years before this and dominated most terrestrial niches. Clearly, in no way were mammals out-competing dinosaurs as a general rule—if anything, it was the other way around, as dinosaurs had reigned supreme over the inconsequential ratlike ancestors of mammals. It was only the chance event of the impact, driven by celestial mechanics independent of the happenings on Earth, which, by wiping out the dinosaurs, opened the way for mammals to thrive and helped establish the world in which we live. In contemplating the extinction of dinosaurs, as it relates to the prospects for the longevity of our own species, Gould notes: "*Dinosaur* should be a term of praise, not of opprobrium. They reigned for 100 million years and died through no fault of their own; *Homo sapiens* is nowhere near a million years old and has limited prospects, entirely self-imposed, for extended geological longevity."[5]

Similarly, trilobites, which were one of the most common creatures in the Paleozoic era, were wiped out in the still unexplained mass extinction that occurred at the end of the Permian period, 250 million years ago.[6] Before this great extinction, which plowed under well over 90 percent of all animal species then extant, there was no sign of trilobites waning—indeed, they appeared to be a remarkably successful lineage. Evolution clearly has no direction or ultimate purpose. Rather, evolution wanders across a terrain of contingent pathways, constrained by the day-to-day demands of natural selection and periodically jostled by unforeseen events.

Once the fundamental importance of contingency is properly appreciated, we are left with the deep recognition that humans were not somehow meant to exist, that evolution does not "progress," and that there is no general tendency for evolution to produce intelligent beings such as ourselves.

Developing the generality of his arguments about the nonprogressive nature of evolution, Gould takes on the emergence of complex life-forms in his book *Full House* (the British edition is titled *Life's Grandeur*).[7] He notes how people commonly assume that evolution has a tendency toward producing more complex creatures, because the most complex creature of any particular era tends to be more complex than the most complex creature from a previous era. Gould reveals how this misconception is rooted in a misunderstanding of statistical processes. He explains that, emerging from chemical constituents through the physics of self-organizing processes, life necessarily started at the simplest level: the humble single-celled organism.[8] After its origin, life-forms were free to vary in many ways. However, by a point of definition they did not have room to become much less complex, since to do so would mean they would no longer be alive and, therefore, would not be counted in our assessment of life-forms. In adapting to their local environments, there was not necessarily a universal drive to complexity, but since chance variance on the simplicity-complexity continuum was censored on the simplicity end, variance could in effect only go in one direction. Thus, although simple single-celled organisms remain by far the most common form of life, the further we move in time from the origin of life to the present, the more opportunity there has been for chance variance in the direction of increased complexity to occur. This creates the pattern where the most complex life-form present at anytime is typically more complex the closer we are to the present.

To retain a historical and biological perspective, Gould emphasizes that throughout all of history, the humble bacteria has always been the most common life-form by any meaningful measure—i.e., the typical life-form has become no more complex over time. It is important to note that by saying increases in complexity occur by chance, Gould is not arguing that evolution itself is a process of chance—a point of confusion among intelligent design supporters. Natural selection is very definitely an organizing force. Organisms have good "design" because of natural selection (and structural forces, as discussed in the previous chapter). However, there is no general reason why it is advantageous to be more or less complex in any particular environment. Thus, although in any specific instance the development of a complex feature may be driven by natural selection, there is no overarching push for complexity in general. That is to say, once we move away from the simplest end of the complexity continuum, any particular species is just as likely to become less complex as more so at any given time as it adapts to its particular environment. For example, many species that took up the role of parasites became less complex as they adapted to and created the parasite niche.[9] Thus, there is no general drive toward complexity. Highly complex organisms emerge due to a process of diversification in the level of complexity rather than a trend in central tendency. Gould makes this point in part to undermine the human arrogance that leads to foolish ideas that the world was made for us and that our emergence was the point toward which all history has striven.

The famous evolutionary theorist (and socialist) J. B. S. Haldane is (perhaps apocryphally) said to have once responded to the question, "What has the study of the creation taught you about the Creator?" with the quip "He has an inordinate fondness for beetles." This comment reflects perhaps one of the most striking observations in natural history: well over a quarter of the million-plus species of animals that have been identified are beetles (whereas less than one-half of 1 percent are mammals). Clearly, if there is a creator, His creation is far more about beetles than about humans. Extending the point even further, Gould notes that we have never lived in the "Age of Man" as commonly professed. We are now and always have been in the "Age of Bacteria."[10]

Emergence in the Dialectical World

Nature is often viewed in simplistic terms: either it is designed for a specific purpose or it is determined by an inherent order. In this, our understanding of the world is imprisoned, as Epicurus noted, by the bonds of fate and determinism. As a result, nature is imbued with progressive properties and an ultimate purpose that undermines inquiry into the physical world. This approach also misses the grandeur of natural history. It proposes a retreat to the confines of faith, a closed system of revealed truth, opposed to open inquiry about the natural world.[11]

Evolution is not an unfolding process with predictable outcomes, but a contingent, wandering pathway through a material world of constraints and possibilities. The contingent, improbable character of evolution is for Gould evidence that it is history (whether natural or social) that is the real designer, the real force behind how the world is organized. It is misleading to look for ideal design, whether as proof of evolution or as divine conception. As Gould explains: "Our textbooks like to illustrate evolution with examples of optimal design—nearly perfect mimicry of a dead leaf by a butterfly or of a poisonous species by a palatable relative. But ideal design is a lousy argument for evolution, for it mimics the postulated action of an omnipotent creator. Odd arrangements and funny solutions are the proof of evolution—paths that a sensible God would never tread but that a natural process, constrained by history, follows perforce."[12]

In replying directly to Gould's criticism in *The Panda's Thumb* that "odd arrangements and funny solutions" point to the reality of evolution, intelligent design proponent William Dembski outdoes himself in launching one sortie after another on Gould's position. First of all, "the design theorist," he tells us, "is not committed to every biological structure being designed." Some, even most (if not quite all), can be evolutionary adaptations. Second, intelligent design should not be confused with "optimal design." The notion of optimal design, Dembski claims, robs design of all "practical significance." Hence, an Intelligent Designer could produce "suboptimal design"—the kind of quirky results pointed to by Gould. Third, "not knowing the objectives of the designer, Gould is in no position to say whether the designer has come up with a faulty compromise among those [various conflicting] objectives." Fourth, Gould like others—such as Lucretius and Darwin before him, who had pointed to malformations or cruelty in nature as indicating the absence of divine providence—conflates science with theology by ignoring the problem of evil. There is no valid reason, Dembski states, to assume "a God who is omnipotent and benevolent in the face of evil." Hence, intelligent design could take the form of a "torture chamber." Indeed, "a torture chamber replete with instruments of torture is designed, and the evil of its designer does nothing to undercut the torture chamber's design." However, this is not, we are informed, a problem that science itself can address, since "the problem of evil is a theological problem," not a scientific one.

By the end of this enormously convoluted series of attempts to refute Gould's position by every means he can think of, Dembski has undercut still further intelligent design's already extremely dubious claims to scientific argument. Design, we are informed, is "actual," but apparently it is to be inferred only in rare cases. Intelligent design is normally suboptimal (making it hard to detect what is designed). It is not directly recognizable to human reason, which cannot apprehend the full purposes of the designer. It can manifest evil as well as good and hence can lead to irrational and horrible results. Indeed, as Dembski exhorts: "This is a fallen world. The good that God initially intended is no longer fully in evidence. . . . More often than we would like design has gotten perverted." Only theology can "discern God's hand in creation despite the occlusions of evil." Teleology itself is subverted by its alter-ego or "dysteleology," that is, the purposiveness of evil.

What emerges from this type of convoluted argument is something that appears to confirm the wisdom of the early Church father Tertullian in declaring that faith is the acceptance of the absurd. Such a call for the acceptance of absurdity on the grounds of faith is the only possible "rational" defense of a creationism that seeks to reduce nature to divine teleology (including "dysteleology"). It provides no basis to understand or investigate the world and its history. An evolutionary perspective, in contrast, since it rejects design, has no need for the concept of "a fallen world" to explain why "God's design" has frequently been "perverted."[13]

Evolutionary theory at its best focuses on the complex and contingent paths that characterize nature in all its diversity. A historical and materialist approach to evolution, emphasizing a dialectical view that focuses on constantly changing relations and processes, offers powerful insights into the emergence and evolution of life, as well as its complexity. Thus Richard Levins and Richard Lewontin contend that the larger, physical world in which all life is situated is filled with its own contingent history and structural conditions. In other words, the world "is constantly in motion. Constraints become variables, causes become effects, and systems develop, destroying the conditions that gave rise to them." The universe is one of change due to existing and evolving tensions, which force transformation in the conditions of the world due to "the actions of opposing forces on them, and things are the way they are because of the temporary balance of opposing forces."[14]

This approach builds on evolutionary theory, embracing chance and necessity, contingency and emergence. In this, natural history is situated within a world consisting of a multitude of forces and pathways. As noted in the previous chapter, structures influence, in part, the course of natural selection, while at the same time the processes of life transform them. Understanding this, the organism is a site of interaction between the environment and genes. In this dialectical relationship the organism and environment exist together, in tension, given that the organism is part of nature. The former is dependent upon the latter for its existence, and both realms are transformed throughout their relationship, but "do not completely determine each other."[15]

At the center of Levins and Lewontin's analysis is a focus on interaction, transformation, and historical constraints. Life is not simply a free-flowing, hodgepodge series of independent events. Instead, it emerges from the complex interactions that are constantly taking place. An organism is both a subject and object, and a dialectical approach illuminates the interaction between organisms and the environment. In day-to-day operations, any number of materials (rocks, water, etc.) exist in the environment, but organisms interact and utilize a small portion of what is available; thus in their patterns of life they determine what is relevant to their development. In the process of obtaining sustenance, organisms must interact with their environment, and in so doing they transform the external world—both for themselves and other species. Their consumption of parts of the external world is also the production of new environments. Of course, the conditions of the environment are not wholly of organisms' own choosing, given that there are natural processes independent of a particular species. Previously living agents have historically shaped nature, and coexisting species are also engaged in altering material conditions.

The "traits" selected in evolution are influenced by the dynamic organism-environment relationship. "Neither trait nor environment exist independently," thus what becomes useful is a consequence of a long historical process—one that is subject to change. An organism is the result of complex interactions between its genes and environment, as well as its own structure, where the organism takes part in the creation of its environment, its ecological niche, and its own construction. In this, it sets—in part—the conditions of its natural selection, by being both the object and subject, in a world indifferent to the success of any given species.

Multiple pathways or channels exist, in relation to the structural integrity (organization) of organisms, for evolutionary processes—in fact, they are part of what created life and makes its continuance possible. Even when the external conditions are fixed, multiple pathways exist, as organisms interact with opposing forces while obtaining the needed materials for survival. What survives is not necessarily due to inherent superiority, but has much to do with contingent events given the multitude of influences that shape the world.

Both emergence and contingency are foundational concepts for analyzing a dynamic world. Change is the rule of life. Organic processes, since they are historically contingent, defy rigid universal explanations. Both the parameters of change and the nature of transformation are subject to change given the ongoing development of life.[16] In such a materialist-dialectical view the notion of "intelligent design" (and an Intelligent Designer) is superfluous, necessarily empty of all genuine scientific content. It is a dead concept meant to displace reason with the *deus ex machina* of an omnipotent God. Instead it promotes a notion of God that is itself redundant.

*

The world of the present, in both its social and natural aspects, is only one of the many worlds that are possible. Gould makes this point with a powerful metaphor.[17] He argues that due to the innumerable contingencies that shaped life's history, if we were to "replay the tape of life" a different history would unfold, one almost certainly without the appearance of humans or any creature much like us. Such an alternative history would appear as sensible and "inevitable" as the history of the world in which we live. This argument challenges intelligent design on the most fundamental level, because it even denies theists refuge in a God who designs only at the start rather than intervening subsequently. The history of life does not suggest the unfolding of a plan. It does not represent a foreordained order. Just as we make our own history—as Marx claimed—nature makes her own history as well.

10. The End of the Wedge

Both Darwin's evolutionary theory and intelligent design creationism have relied, but in entirely different ways, on the metaphor of a wedge. In conceiving his theory of natural selection Darwin introduced a powerful metaphor of a log or other solid area crammed with wedges, some of which were being driven home, thereby forcing out others. "Nature," he wrote,

> may be compared to a surface covered with ten thousand sharp wedges, many of the same shape, and many of different shapes representing different species, all packed closely together and all driven in by incessant blows: the blows being far severer at one time than at another; sometimes a wedge of one form and sometimes another being struck; the one driven deeply in forcing out others; with the jar and shock often transmitted very far to other wedges in many lines of direction.[1]

The wedge metaphor had originally been introduced in Darwin's notebooks at the very time (September 1838) at which, under the influence of Malthus, he had discovered his theory of natural selection. Darwin subsequently worked on perfecting it in his later (unpublished) book *Natural Selection*, and employed it as well, in a more compressed version, in the "abstract" that was to become the *Origin of Species*. As Stephen Jay Gould explained:

Nature, Darwin believed, is full of species ("a surface covered with ten thousand wedges…all packed closely together"). All potential addresses are occupied, but new challengers continually arrive to compete for space. They can succeed in a full world only by driving other species out in overt competition for limited resources ("the one driven deeply in forcing out others").[2]

Given the fame of Darwin's wedge metaphor, which came to stand for the force of natural selection, it is curious and ironic that intelligent design proponents have themselves adopted a wedge metaphor—albeit of a sharply different character. As Phillip E. Johnson put it: "A log is a seeming solid object, but a wedge can eventually split it by penetrating a crack and gradually widening the split. In this case the ideology of scientific materialism is the apparently solid log."[3] Here the aspect of competition between different wedges (species) is missing, and instead the goal is the destruction of the environment (the log) itself and all that goes with it. Rather than Darwin's wedge of natural selection, the aim is the artificial wedging in of a creationist theological and moral order at the expense of the entire scientific-social-cultural environment. The term *wedge*, according to William Dembski, "has come to denote an intellectual and cultural movement," a definite "strategy for unseating materialism and evolution."[4]

The single wedge of intelligent design is viewed here as the wedge of God or Logos, directed at materialism and evolutionary science. The metaphor takes its significance in this case from the fact that though the thin end of the wedge is intelligent design dressed up as science (and is referred to by intelligent design proponents as "scientific renewal"), the thick end of the wedge is fundamentalist Christian theology and morality (or "cultural renewal").[5] It is the hammering in of the wedge to the point that the thick end enters in, splitting the material surface below that intelligent design proponents see as its object. The fact that the image is almost exclusively one of destruction (the splitting of the log of materialism) fits well with a fundamentalist Christian eschatology of the end of the world and the Second Coming.

In his book *The Wedge of Truth* (as noted in chapter 2) Johnson made the far-reaching objectives of the intelligent design movement's wedge strategy explicit:

> The Wedge of my title is an informal movement of like-minded thinkers in which I have taken a leading role. Our strategy is to drive the thin edge of our Wedge into the cracks in the log of naturalism by bringing long-neglected questions to the surface and introducing them into public debate. Of course the initial penetration is not the whole story because the Wedge can split the log only if it thickens as it penetrates.[6]

The wedge of the intelligent design movement thus has a thick and a thin end. At the thin end intelligent design is deliberately introduced in the guise of a "scientific" theory to counter Darwinism. This initially takes the form of an assault on science education in public schools, as marked by the 2005 trial in Dover, Pennsylvania, where intelligent design creationism was presented as an alternative to Darwinian evolution. As the wedge is forced in, it widens and becomes a larger critique of materialism in the realms of society, culture, and politics, giving Christian theology an expanded moral role in society. The leading figures of materialism—namely Darwin, Marx, and Freud (and before them Epicurus)—are to be displaced by the new prophets of creationism.

This was blatantly expressed in the 1999 *Wedge Document* of the Discovery Institute's Center for the Renewal of Science and Culture, which explained that the ultimate goal of "cultural renewal" was to ensure that design theory penetrated the social sciences, humanities, and culture in general, namely the disciplines of "psychology, ethics, politics, theology, and philosophy in the humanities." This included plans to alter contemporary views of "abortion, sexuality and belief in God." To achieve this it was necessary to undermine the views of materialist "thinkers such as Charles

Darwin, Karl Marx, and Sigmund Freud" who "portrayed humans not as moral and spiritual beings, but as animals or machines who inhabited a universe ruled by purely impersonal forces."[7]

The intent of the intelligent design movement is therefore only superficially to counter evolution and to present an alternative "science." Rather, as its own documents make clear, its deeper purpose is to demonstrate that "design points to a knowable moral order" ruled over by the Intelligent Designer.[8] If Darwin used the wedge metaphor to explicate the role of natural selection in nature (thereby reinforcing materialist views of the world), today's intelligent design proponents use it as a device to split materialism in culture as well as science, and to open the way once again to God's dominion.

The Thick End of the Wedge I: Theology

Intelligent design proponents present their analysis as science, not theology. Yet their intent is not to advance science, but to use scientific claims as a wedge for theological arguments. This was clearly demonstrated in William Dembski's 1999 book *Intelligent Design: The Bridge Between Science and Theology*. Dembski sought throughout his book to draw a sharp distinction between intelligent design and creationism. "The most obvious difference between the two," he wrote, "is that scientific creationism has prior religious commitments whereas intelligent design does not." Intelligent design has less religious "baggage." It is, Dembski claimed, compatible with a much larger playing field than Christian creationism, as much at home with Plato and the Stoics as with young-earth creationism. "Intelligent design presupposes neither a creator nor miracles. Intelligent design is theologically minimalist. It detects intelligence without speculating about the nature of the intelligence." Dembski asserted that British natural theology of the seventeenth to nineteenth centuries was a precursor to today's intelligent design perspectives, but made the mistake of mixing science with theology, deriving conclusions about the nature of God.

The new intelligent design movement, he contended, is free of such errors, creating a firewall between science and theology. It provided a bridge to the latter without itself crossing the bridge, remaining on the grounds of science.[9]

Nevertheless, Dembski, as one of the principal proponents of intelligent design, could be seen in the same book and often in the same chapters repeatedly crossing the "bridge" from theology to science and science to theology. "Intelligent design," he stated at one point, "is just the Logos theology of John's Gospel restated in the idiom of information theory."[10] The Intelligent Designer in the form of God, Christ, and Logos permeates his work. References to design in nature are accompanied by references to scripture. Science is seen as incomplete and "deficient" without Christ. A series of quotations from Dembski's *Intelligent Design* will suffice to show the bridge between "science" and theology promulgated by his thought:

- The temptation to worship and serve the creature rather than the creator is ever present to us. It happens when we lose sight that this is God's world and that nothing happens apart from his consent.

- If we take seriously the word-flesh Christology of Chalcedon (i.e. the doctrine that Christ is fully human and fully divine) and view Christ as the *telos* toward which God is drawing the whole of creation, then any view of the sciences that leaves Christ out of the picture must be seen as fundamentally deficient.

- Christ is also the incarnate Word who through the incarnation enters and transforms the whole of reality.

- The point to understand is that Christ is never an *addendum* to a scientific theory but always a *completion*.

- Christology tells us that the conceptual soundness of a scientific theory cannot be maintained apart from Christ.... Christ is the light and the life of the world. All things were created by him and for him. Christ defines humanity, the world and its destiny. It follows that a scientist, in trying to understand some aspect of the world, is in the first instance concerned with that aspect as it relates to Christ—and this is true regardless of whether the scientist acknowledges Christ.

- Christ is indispensable to any scientific theory, even if its practitioners don't have a clue about him. . . . [T]he conceptual soundness of the theory can in the end only be located in Christ.

- The language that proceeds from God's mouth in the act of creation is not some linguistic convention. Rather as John's Gospel informs us, it is the divine *Logos*, the Word that in Christ was made flesh and through whom all things are created.

- Theology gets its data from Scripture, science from nature. Nature may therefore testify to God in ways quite distinct from Scripture.

- To say that God through the divine *Logos* acts as an intelligent agent to create the world is only half the story. . . . In addition, the world is intelligible. . . . God, in speaking the divine *Logos*, not only creates the world but also renders it intelligible.

- Theology has traditionally been called "the queen of the sciences."[11]

In this way Dembski makes clear that the end of the argument from intelligent design is to provide a bridge from theology, which derives its knowledge of God through scripture, to science—and from science back again to theology. In this respect, intelligent design has the same objectives, despite his protestations, as the British natural theology of the seventeenth to the nineteenth centuries. Indeed, nowhere in Paley is the connection between nature and scripture made so evident.

Intelligent design proponents thus reject the notion of a conflict between science and theology, and also the alternative view that science and religion can be treated as different compartmentalized areas of human interest (such as Stephen Jay Gould's NOMA, or "non-overlapping magisteria" of science and religion). Rather, they portray science and theology as "mutually supportive," in Dembski's words. For example, "the theologian may learn from the physicist that the universe began as an infinitely dense fireball known as the Big Bang, whereas the physicist may learn from the theologian that God created the world by means of a divine logos." Intelligent design, we are told, merely points to design, giving the

Designer a foothold in the world of science, whereas the task of theology is to connect this to the God of scripture. But the God of scripture also informs and constrains science. As Phillip E. Johnson put it, the problem with Darwinism is its conflict with Christian scripture: "the Darwinian theory of evolution" is unacceptable because it "contradicts not just the Book of Genesis, but every word in the Bible from beginning to end. . . . In the beginning was the word. In the beginning was intelligence, purpose, and wisdom. The Bible had that right. And the materialist scientists are deluding themselves."[12]

Hence, it is affirmation of Christ and his effects rather than the development of science that constitutes the main purpose of intelligent design creationism—as propounded by its main advocates. The Discovery Institute's 1999 *Wedge Document* insisted on the need to replace materialist science with "a science consonant with Christian and theistic convictions." According to Jay Wesley Richards, writing in 2001 in *Signs of Intelligence* (co-edited by Dembski and James Kushiner), intelligent design "is a valuable resource for Christian apologetics. Positively, not only can intelligent design become—by extension—an apologetic argument, but it also proposes a view of natural science compatible with the Christian doctrine of creation."[13]

Although proponents of intelligent design claim the logical independence of their design argument from religious presuppositions, they frequently fall back on the notion that the inconsistencies of their views can only be overcome, not within the domain of science, but within the domain of theology, which supports and completes science. In this sense the thin wedge of intelligent design within science requires the thick edge of theology for its completion.

This dependence of intelligent design creationism on theology to complete its argument is clearest where attempts are made to reply to the age-old criticism of the argument from design—associated with thinkers like Epicurus, Lucretius, Hume, Darwin, and

Gould—that the world exhibits not so much instances of optimal design, as suboptimal design. Nature gives rise to all sorts of quirky solutions, and real torments and disasters (ants, as Darwin pointed out, that have slaves; floods that kill thousands). How then can this be inferred as *intelligent* design? In responding Dembski claims, on the one hand, that design is necessarily "imperfect" and "constrained" (even if intelligent), and, on the other, that such portrayals of suboptimal design are often pointing to the problem of evil. "Critics who invoke the problem of evil against intelligent design have left science behind and are engaging in philosophy and theology. . . . Dysteleology, the perversion of design in nature, is a reality." Not only does he hint here of a greater, intelligent evil (the Devil?), but we are informed that "this is a fallen world. The good that God initially intended is no longer fully in evidence." To address this problem of evil, Dembski argues, theology rather than science is required. Theology's "task is to focus on the light of God's truth and thereby dispel evil's shadows." For Benjamin Wiker and Jonathan Witt, too, the problem of "bad design" is associated with the problem of "evil." Consequently intelligent design requires the thick end of the theological wedge to complete its argument, which is not self-sufficient (nor is it intended to be) within the domain of science: either with respect to its teleology or its dysteleology.[14]

The Thick End of the Wedge II:
Politics, Morality, and Culture

Evangelical Christianity in the United States is a political as well as a religious movement. The attack on Darwinian evolution—the thin end of the wedge of the intelligent design movement—has understandably attracted most of the attention thus far. But the intelligent design strategy derives its ultimate meaning from the hammering in of the thick end of the wedge, represented not only by creationism as such, but also by an ultraconservative political thrust that lies at the heart of Christian evangelicalism. It is this conservative culture, morality, and politics that the wedge of intelligent design is principally meant to force into society, displacing its chief enemy in the form of materialist science and philosophy.

This can be clearly seen in the writings of leading intelligent design figures, who have taken on the wedge strategy's task of "cultural renewal." A noted case is political scientist John G. West, author of the opening paragraphs of the *Wedge Document* and associate director of the Discovery Institute's Center for Science and Culture. In his 2007 book, *Darwin Day in America: How Our Politics and Culture Have Been Dehumanized in the Name of Science*, West depicted the contemporary "dehumanization" of all aspects of criminal justice, social welfare, public education, human sexuality, and life and death as the direct result of the rise of scientific materialism. He traced materialism to the Greek atomists, Leucippus, Democritus, and Epicurus, and the Roman poet Lucretius, whom he claimed "saw human beings and the rest of nature as products of the mindless collision of atoms."[15] These materialist ideas reappeared in the scientific revolution of the seventeenth and eighteenth centuries, and indeed were instrumental in bringing about modern science and subsequently gaining their most important nineteenth-century scientific adherent in Charles Darwin. "Darwin helped transform materialism from a fantastic tale told by a few thinkers on the fringe of society to a hallowed scientific principle enshrined at the heart of modern science. . . . Darwin [thus] helped spark an intellectual revolution that sought to apply materialism to nearly every area of human endeavor. This new thoroughly 'scientific' materialism affected the entire span of culture, from economics and politics to education and the arts."

West's first two chapters critically exploring the rise of materialism were supplemented later by a chapter attacking social Darwinism (in which George W. Bush was praised, along with George Gilder and Newt Gingrich, as an anti-social Darwinist, and socialists were presented as consummate social Darwinists) and by two chapters defending intelligent design creationism in the context of current controversies. Among his accusations was the claim that evolutionary theory had been used to promote gay rights in science classrooms.[16]

The larger purpose of West's analysis, however, was to upend the materialist bases of contemporary philosophy and culture, arguing that under its corrupting influence (1) criminal justice denied the existence of free will on the part of the criminal, "turning punishment into treatment"; (2) welfare emphasized eugenics and other "scientific" techniques rather than moral principles; (3) the "science of business" viewed human beings as subjects for manipulation through advertising; (4) sex education in the schools had been used to promote free sexual activity, and to attack abstinence education; and (5) the Christian sanctity of life had been transformed into a culture of death through the promotion of abortion and euthanasia. *Darwin Day in America* concluded with the argument that the renewal of science under intelligent design "will likely have as dramatic an impact on the politics and culture of the future as scientific materialism has had on the politics and culture of the past."

Intelligent design was thus seen as the thin end of the wedge; and with the further hammering in of the wedge there would be a "cultural renewal" along Christian evangelical lines.[17] "[F]or the first time since the materialist onslaught," thanks to the intelligent design movement, West wrote in his article "C. S. Lewis and Materialism," "we have an opportunity to bring about the collapse of materialism and to re-found both science and culture along the [Christian] lines envisioned by C. S. Lewis more than half-a-century ago," including the overturning of the "modern welfare state" and "materialist social science."[18]

In his *Moral Darwinism: How We Became Hedonists*, Wiker argued at length that modern materialism, including Darwinism, grew out of Epicurean materialism, which was "designed to destroy all religion" and was long viewed as the chief nemesis. It was on Epicurus, and after Epicurus on Darwin, Marx, and Freud, that the blame was to be laid for materialism's "larger package" of "Social Darwinism and eugenics. . . . libertinism, abortion, infanticide, euthanasia, cloning"—in short, "hedonism."

In *Architects of the Culture of Death*, Donald De Marco and Wiker pronounced that materialism was a "culture of death" opposed to the "culture of life" of Christianity. Separate chapters were devoted to attacking Karl Marx, Charles Darwin, Sigmund Freud, Auguste Comte, Jean-Paul Sartre, Simon de Beauvoir, Wilhelm Reich, Margaret Mead, Margaret Sanger, and Jack Kevorkian for "the self-willed eclipse of the true sense of God and man that defines the Culture of Death." In his *Ten Books That Screwed the World, and Five Others That Didn't Help* Wiker dedicated individual chapters to attacks on Niccolò Machiavelli, Thomas Hobbes, Jean Jacques Rousseau, John Stuart Mill, Friedrich Nietzsche, V. I. Lenin, Alfred Kinsey, and Betty Friedan—all of whom were accused of embracing materialism, rejecting Christian morality, and contributing to evil. As in other works by intelligent design proponents, Darwin's evolutionary theory was portrayed as leading to Hitler. In discussing the Nazi Holocaust Wiker stated, "One cannot help but be reminded of Darwin's *Descent of Man*."[19]

The thick end of the wedge with respect to social science, culture, and human civilization equates design with meaning, i.e., God's Logos, and materialism with meaninglessness and even nihilism (if not evil itself). For Wiker and Witt, "Epicureanism provided the prototype of the meaningless universe—godless, governed by chance, purposeless." Over the course of history, they argued, "many springs and rivulets" fed "the great river of materialism/relativism/nihilism, but since the Victorian era, its principal tributaries are Freudianism, Marxism, and, above these, Darwinism." Freud reduced everything "to two motives[:] the fear of death and the desire for sex." Marx reduced everything "to the pushings and shovings of human beings variously related to the material means of production, caught and defined in every thought and deed by the shaping force of class struggle." And Darwin reduced all to the animalist "desire to survive and propagate." None of the great materialist thinkers, we are told, left room for meaning or Christian morality.[20]

Nietzsche's nihilism was depicted by these thinkers as a logical outgrowth of such materialism as was Sartre's existentialism and Derrida's extreme relativism and deconstruction. Sartre, Nietzsche, and Derrida were accused of "suffocating [culture] in the materialist darkness, where all meanings are mere human fabrications" as opposed to Divine Logos. Derrida was specifically criticized for having advanced "a view of language that is pure *misère*, unmitigated nihilistic darkness, a language of unmeaning fit for a meaningless world. In this, Derrida has inadvertently done us an invaluable service . . . for he has traced out the implications for meaning in a world without God: by removing the Author, the materialists created a meaningless drama." Indeed, "when modernity adopted Epicurean materialism as its scientific foundation and reality filter," Wiker stated in *Moral Darwinism*,

> it simply reinstated the ancient belief in the amorality of nature. The intrinsic purposefulness of nature, which was the foundation of moral claims according to the Christian natural law argument, was given the *coup de grâce* by Darwin. . . . Whatever a particular materialist may happen to desire morally, it is simply an incontrovertible fact that, with the increasing secularization of the West, the repugnance toward abortion, infanticide, eugenics, euthanasia and sexual libertinism, which had its theoretical and historical origin in Christianity (stretching back through Judaism), has given way to acceptance. The cause for this moral reversal is secularization, and as we have seen, the cause of secularization has been the rise of Epicurean materialism as culminating in moral Darwinism.[21]

Counter to this was the "meaningful world" of Logos, which intelligent design proponents present as the social counterpart of design within nature. "The Divine Logos....identifying logos as the Son," Wiker and Witt wrote in *The Meaningful World*, "is particular to Christianity; but the understanding that nature revealed the logos—was not. It was found, in one form or another, among many Greek and Roman philosophers, in particular the Stoics." The goal of materialism, they stated, was to "remove God and enthrone chance," thereby removing "the reality of both good and evil," i.e., God's Logos. Yet, without God neither meaning nor morality was possible.[22]

God as Superfluous to Science and Human Freedom

Epicurus, in Marx's words, insisted that "the world must be *disillusioned*, and especially freed from fear of gods, for the world is my *friend*."[23] The gods were seen as having no relation to the material world, and humanity was freed from the bonds of fate to confront the physical world and their own freedom and mortality. For this reason, Epicurus's philosophy was viewed as the chief enemy of the argument from design first introduced by Socrates, and later adopted by Plato, the Stoics, and early Christianity. Epicurus was the chief enemy not only of Christian teleology but also of the world alienation characteristic of Christianity from the time of the early Church founders.

In sharp contrast to "the world is my *friend*" of Epicurus, Christianity historically embraced the New Testament's Epistle of St. James: "Whosoever . . . desireth to be a friend of the world is an enemy of God." Indeed, the world was characterized as "earthy, animal, devilish." As Helen Ellerbe has noted in a discussion of Christianity's "Alienation from Nature," "Nature was . . . seen as the realm of the devil. The Church chose the image of Pan, the Greek god of nature, to portray the devil. The horned, hoofed, and goat-legged man had been associated with a number of fertility figures and had previously been deemed essential to rural well-being. . . . His name, 'Pan,' meant 'all' and 'bread.' But, particularly after the turn of the millennium when the Church authorized specific portrayals of the devil, the vilified Pan came to evoke terror or 'panic' as the image of Satan."[24]

This deep antipathy to the material world characterizing Christian thought was exemplified in the mid-twentieth century by the writings of C. S. Lewis, who has since been adopted as the patron saint of intelligent design proponents. Lewis's anti-materialism/anti-naturalism was most evident in his 1947 *Miracles*, where he declared:

> Just because the Naturalist thinks nothing but Nature exists, the word *Nature* means to him merely "everything" or "the whole show" or "whatever there is." And if that is what we mean by Nature, then of course nothing else exists. . . . If there is nothing but Nature, therefore, reason must have come into existence by a historical process. And of course, for the Naturalist, this process was not designed to produce a mental behavior that can find truth. There was no Designer.[25]

In *The Abolition of Man*, Lewis suggested that to view human beings primarily in materialist terms was tantamount to the "abolition" of humanity, i.e., dehumanization and loss of moral and spiritual certitude. In his fictional space trilogy—*Out of the Silent Planet*, *Perelandra*, and *That Hideous Strength*—he carried this further, characterizing the planet Earth as the "silent planet" (removed from Logos) and ruled by Satan and his materialist followers. The planets Mars and Venus, in contrast, were the realm of angels. The "silent planet" was a fallen world, awaiting regeneration through Christ. Hence, to seek its material transformation was a dangerous, utopian (even Satanic) delusion. In "Funeral of a Great Myth," Lewis raised questions about the validity of evolution (at least in its more mythological proportions), which he saw as mainly impressing the "folk imagination" with the myth of progress. God was an absolute necessity for the world. The concept of evolution was doubly suspect because it arose in the "Revolutionary period" of modern politics. In "Two Lectures," Lewis suggested that just as a rocket has a designer, so must nature. At another point in his work he wrote: "If the universe is not governed by absolute goodness [a benevolent God] then all our efforts are in the long run hopeless."[26]

Such hopelessness in the absence of a benevolent deity, however, is obviously not characteristic of materialism, which places its hope in *homo faber*. (When God was dethroned by materialism, Lewis complains, "Man ascended his throne. Man has become God.")[27] What is truly objectionable about the materialist bases of science from the intelligent design standpoint is not the rejection of God (atheism), which is hardly ever the point, but that materialism removes the earthy need for God, who becomes superfluous in accounting for the world and human freedom.

Indeed, scarcely less objectionable than outright materialism from an intelligent design standpoint is the work of American pragmatists like Charles Peirce and William James, for whom God became something of a pragmatic, psychological necessity—but one that had no direct relation to the material world or science. As Peirce (who counted Epicurus among his major influences) once noted: "To the mind of a physicist there ought to be a strong presumption against every mystical theory; and therefore it seems to me that those scientific men who have sought to make out that science was not hostile to theology have not been so clear-sighted as their opponents." For Peirce the only religion easily tolerated by science was that which propounded a deity in the form of an abstract "Supreme Ideal," such that it was "repugnant to its real existence." In other words, such a deity would have to be superfluous to the material world, which must be understood on a purely material basis. The turning of the entire magisterium of nature over to materialism did not in Peirce's view eliminate the possibility of a religious morality or belief in God. But he argued, like Epicurus, Darwin, Marx, and Gould, that God (or the gods) had no connection to the magisterium of science, which encompassed all of worldly reality. Dialectical materialists, such as Marx, went one step further, arguing that God was not merely superfluous but a manifestation of an inverted world, and was thus an alienated human product.[28]

All of this helps us understand more fully the extremely virulent attacks by intelligent design proponents on all varieties of materialist thought, most notably Epicurus, and the modern unholy trinity of Darwin, Marx, and Freud. Phillip E. Johnson rails against Spinoza's God, Einstein's God, and Hawking's God as mere abstractions, since the material world has been given over entirely to materialism. For Johnson, Gould's NOMA is nothing more than a "power play" that "bars religion from claiming that there is a supernatural creator (much less one who was incarnated in Jesus), a divinely infused soul, a life after physical death or a source of divine revelation such as inspired Scripture. This is 'separate but equal' [of the magisteria] of the *apartheid* variety." God is left with "no cognitive status." "Accommodation" with scientific materialism, he adds, "doesn't work since religion is acceptable to materialists only as long as it stays in the realm of the imagination and makes no independent claims about objective reality."[29]

Likewise Wiker claims that the materialist approach, even when it aims at a kind of perpetual cease-fire, as in Gould's NOMA, gives to religion's magisterium the "morality of morals" but insists that the "anthropology of morals" belongs to science. This, however, is a deception since materialists from Epicurus to the present (including Darwin, Marx, and Freud) have sought to reduce all morality to the anthropology of morals, discounting foundationalism and hence God's intelligent design of the moral world. As Wiker puts it, " 'factual conclusions' about nature entail, of necessity, that these conclusions be applied to human nature, and that means materialist science cannot and will not honor the terms of this false peace."[30] Of course, the point of Gould's distinction is to emphasize that, just as there are no divine answers to moral questions, nature does not provide us with such answers either. Questions of what we ought to do (the morality of morals) are indeed important, but they do not have absolute answers—they can only be answered by people in the context of their times.

Natural scientists, along with their social science counterparts, typically reject arguments that suggest the world is predetermined, teleological, or governed by miracles and divine intervention. The resurgence of intelligent design is thus nothing less than an attempt by theistic thinkers (predominantly fundamentalist Christians) to reclaim a hold in the material world, from which they (and their God) were largely excluded following the Enlightenment and the Darwinian revolution. Intelligent design is thus first and foremost an attack on materialism-humanism and its conception of historical emergence in the natural and social world.

This struggle between materialism and creationism (intelligent design) has now lasted for thousands of years, from ancient Greece to the present. Like the natural theologians of centuries past, the claim of today's intelligent design proponents is that they can provide evidence of design (or Logos) in nature that supplements revelation/scripture. Counter to Marx's critique of heaven as the basis for a critique of earth, intelligent design offers a teleology of earth (natural and social) as the proof of a teleology of heaven.

Sociologist of science and intelligent design defender Steve Fuller has gone so far as to declare on quasi-pragmatic grounds (though far removed from American pragmatism) that divine teleology is a superior ground for science. Accusing Darwinism of narrow adherence to materialism, Fuller asserts that "the general evolutionary perspective that Darwin ultimately championed . . . tended to discourage systematic scientific inquiry, stressing instead the need to cope with our transient material condition in an ultimately pointless reality." In contrast, "Intelligence design theory," he claims, "provides a surer path to a 'progressive' attitude to science than modern evolutionary theory," precisely because of the teleological, anthropocentric, hierarchical views embedded in its "fundamentally religious" perspective, which is focused on "the ennoblement of humanity, the species created in God's image." Fuller asserts that the *belief* in intelligent design produces superior results for science—quite apart from its inherent truth value. Yet Fuller's quasi-pragmatic argument for religious teleology is largely shunned by intelligent design proponents, who claim direct religious truth

for their view, which is aimed not at advancing science but at providing evidence for an Intelligent Designer.[31]

Intelligent design's wedge strategy constantly tries to reduce
our understanding of the world to a choice between the roll of the
dice (pure chance) and God, suggesting that only the latter can
fully explain real world complexity. On the basis that nothing but
God can account for "irreducible complexity," since pure chance
is ruled out as insufficient, we are led to believe that we need to
adopt the whole regressive culture of fundamentalist Christianity:
a servile relation to God, theology (based on biblical scripture and
revelation), and such cultural and political accompaniments as
attacks on homosexuality, abortion, women's rights, welfare, social
planning, democracy, progress, etc.

Materialist and evolutionary approaches to the world, in contrast, argue for the complexity of natural/material processes, i.e.,
contingent evolutionary pathways that are not predetermined and
can be understood only by replaying the tape of life. Here the world
is left open to the full wealth of nature and history, as an ongoing
process of mediation, contradiction, and change. Darwin opposed
"intelligent design" (a term that he was the first to use in its modern sense), by replacing it with the contingent world of "natural
selection," (a term we also owe to him). His wedge metaphor for
natural selection dramatized not only the force but the endless
change and diversity that characterized the natural world. The
world is not a functional, mechanical entity like Paley's watch, but
an endlessly varied process of evolutionary change, without divine
purpose, but not without human-generated historical meaning.

As Stephen Jay Gould wrote of Darwin's evolutionary theory in
Ever Since Darwin:

> I believe that the stumbling block to its acceptance does not lie in any sci-
> entific difficulty, but rather in the radical philosophical content of
> Darwin's message—in its challenge to a set of entrenched Western atti-
> tudes that we are not yet ready to abandon. First, Darwin argues that evo-

lution has no purpose. Individuals struggle to increase the representation of their genes in future generations, and that is all. . . . Second, Darwin maintained that evolution has no direction; it does not lead inevitably to higher things. Organisms become better adapted to their local environments, and that is all. . . . Third, Darwin applied a consistent philosophy of materialism in his interpretation of nature. . . .

Yes, the world has been different ever since Darwin. But no less exciting, instructing, or uplifting; for if we cannot find purpose in nature, we will have to define it for ourselves. . . . I suggest that the true Darwinian spirit might salvage our depleted world by denying a favorite theme of Western arrogance—that we are meant to have control and dominion over the earth and its life because we are the loftiest product of a preordained process.[32]

Likewise Freud insisted that the replacement of the religious *Weltanschauung* by the scientific *Weltanschauung* offered the possibility of liberating humankind from its own illusions and repressions, creating a wider human freedom.

To make space for materialist explanations of society as well as nature, and to advance human freedom, Marx engaged in a critique of heaven and a critique of earth. The critique of heaven, in this view, was a necessary but not sufficient condition for overcoming worldly alienation, which also required for its fulfillment the critique of earth and real-world social transformation. Inspired by Epicurus, Marx emphasized contingency in the natural world, which served as a prerequisite for freedom in the social world. This is why the battle over the natural world was so important. Human society was not abstracted from nature within Marx's dialectical theory. "To say that man's physical and mental life is linked to nature," he wrote, "simply means that nature is linked to itself, for man is a part of nature."[33] Indeed, Marx, like Darwin and Freud, saw the relationship of nature and society as one of coevolution.

Because of this consistent materialism, Marx's historical materialism in particular has remained a crucial social foundation from which to engage in the critique of intelligent design. It resolutely brings a non-mechanistic, non-reductionist, materialist dialectic to the analysis of both nature and society. As Harvard biologist and geneticist Richard Lewontin, building on both Darwin and Marx (if not Freud as well), has written of this uncompromising materialist-scientific viewpoint:

> We take the side of science *in spite* of the patent absurdity of some of its constructs, *in spite* of its failure to fulfill many of its extravagant promises of health and life, *in spite* of the tolerance of the scientific community for unsubstantiated just-so stories, because we have a prior commitment, a commitment to materialism. It is not that the methods and institutions of science somehow compel us to accept a material explanation of the phenomenal world, but, on the contrary, that we are forced by our *a priori* adherence to material causes to create an apparatus of investigation and a set of concepts that produce material explanations. No matter how counter-intuitive, no matter how mystifying to the uninitiated. Moreover, that materialism is absolute, for we cannot allow a Divine foot in the door.[34]

In Marx's view (as in that of Darwin and Freud, Gould and Lewontin), it was crucial to combat all attempts to wedge the "Divine foot" in the natural and physical world. But the same applied as well to the socio-historical world, which is equally a part of the magisterium of materialism-humanism. The first principle of all true science was the overcoming of religious alienation, helping dispel illusion by enhancing human knowledge of the material world. Just as Lucretius wrote "things come into being without the aid of the gods," so for Marx all human history, including the development of human nature and capacities, the formation of new needs, etc., is made by human beings as self-mediating beings of nature, who exist "without the aid of the gods."[35] If there is evidence of design in history, it is because it has a designer—humanity itself, as a result of the unending, historically contingent struggle for development and freedom—in a continual metabolic interchange with the natural world. We can know human history, as Vico said, because we have made it—if not always under conditions of our own choosing. Our relation to nature is also, within limits, a matter of our own choosing.

To turn to "the will of God," in a desperate attempt to account for the world around us, as Spinoza wrote, is to take refuge in "the sanctuary of ignorance." It is to deny human freedom and our responsibility to each other as social beings—along with our responsibility to nature, i.e., life itself. "Philosophy, as long as a drop of blood shall pulse in its world-subduing and absolutely free heart," Marx declared, "will never grow tired of answering its adversaries with the cry of Epicurus: 'Not the man who denies the gods worshipped by the multitude, but he who affirms of the gods what the multitude believes about them, is truly impious.'"[36] Indeed, reason, science, and human freedom only truly commence, as Epicurus recognized in antiquity, once the gods have at last been banished from the earth.

Notes

PREFACE
1. Brett Clark, John Bellamy Foster, and Richard York, "The Critique of Intelligent Design: Epicurus, Marx, Darwin, and Freud and the Materialist Defense of Science," *Theory and Society* 36/6 (2007): 515–46. Various passages throughout this book are drawn from this earlier article, often in modified form.

CHAPTER 1: INTRODUCTION
1. Charles Darwin, *The Correspondence of Charles Darwin* (Cambridge: Cambridge University Press, 1994), 9:135. For a fuller discussion see chapter 6 below.
2. For a critique of the earlier "creation science," see Philip Kitcher, *Abusing Science* (Cambridge, Mass.: MIT Press, 1983); and *Living with Darwin* (New York: Oxford University Press, 2007). In *Living with Darwin*, Kitcher presents a history of creationist explanations of the natural world, noting how these explanations have repeatedly changed, each one being incompatible with the others.
3. Edward J. Larson, *Trial and Error: The American Controversy Over Creation and Evolution* (Oxford: Oxford University Press, 2003), 15–27.
4. Lawrence W. Levine, *Defender of the Faith: William Jennings Bryan* (New York: Oxford University Press, 1965), 277.
5. Edward J. Larson, *Summer for the Gods: The Scopes Trial and America's Continuing Debate* (New York: Basic Books, 1997); Larson, *Trial and Error*, 28–57; Stephen Jay Gould, *Hen's Teeth and Horse's Toes* (New York: W. W. Norton, 1983), 263–79, and *Rocks of Ages: Science and Religion in the Fullness of Life* (New York: Ballantine, 1999), 133–38.
6. Edward Humes, *Monkey Girl: Evolution, Education, Religion, and the Battle for America's Soul* (New York: Ecco, 2007); National Center for Science

Education, "10 Significant Court Decisions Regarding Evolution/ Creationism," http://www.ncseweb.org/resources/articles/5690_10_signifi cant_court_decisions_2_15_2001.asp; John R. Cole, "Wielding the Wedge," in Andrew J. Petto and Laurie R. Godfrey, eds., *Scientists Confront Intelligent Design and Creationism* (New York: W. W. Norton & Company, 2007), 110–128; Michael Shermer, *Why Darwin Matters* (New York: Henry Holt and Company, 2006).

7.   Carl Sagan, *The Demon–Haunted World* (New York: Ballantine Books, 1997), 26–27.

8.   Percival Davis and Dean H. Kenyon (academic editor Charles B. Thaxton), *Of Pandas and People: The Central Question of Biological Origin* (Dallas: Haughton Publishing, 1993), 99–100 (quoted in *Kitzmiller*, see note 10 below). The Discovery Institute was heavily involved in the production of this textbook. Dean Kenyon, one of the two principal authors of *Of Pandas and People*, is a fellow of the Discovery Institute's Center for Science and Culture, as is the academic editor, Charles Thaxton. Stephen C. Meyer and Michael J. Behe, two senior figures in the Center for Science and Culture, are listed in the acknowledgements as "critical reviewers." Nancy R. Pearcey, another fellow at the Center, is listed on the acknowledgement page as an editor and contributor to the overview chapter. Meyer is included under the textbook's "authors and contributors."

9.   Phillip E. Johnson, *Darwin on Trial* (Washington, D.C.: Regnery Gateway, 1991).

10.   United States District Court for the Middle District of Pennsylvania, *Kitzmiller v. Dover Area School District, et al.*, http://www.pamd.uscourts.gov/kitzmiller/ kitzmiller_342.pdf.

11.   Michael Behe, *Darwin's Black Box* (New York: Free Press, 1996), 39; and "Reply to My Critics," *Biology and Philosophy* 16/5 (November 2001), 691, 695; *Kitzmiller*, 73.

12.   Behe, "Reply to My Critics," 695; *Kitzmiller*, 74–78; Kenneth R. Miller, *Only a Theory* (New York: Viking, 2008), 58–62.

13.   *Kitzmiller*, 24–25, 29–31, 35, 70–71, 89; Barbara Forrest and Paul R. Gross, *Creationism's Trojan Horse: The Wedge of Intelligent Design* (Oxford: Oxford University Press, 2007), 325–38; Shermer, *Why Darwin Matters*, 99–105; Humes, *Monkey Girl*; Jon D. Miller, Eugenie C. Scott, and Shinji Okamoto, "Public Acceptance of Evolution in 34 Countries," *Science* 313/5788 (August 11, 2006): 765–66; Alan Sokal, *Beyond the Hoax* (Oxford: Oxford University Press, 2008), 336–40. The term "God of the Gaps" dates back to C. A. Coulson, *Science and Christian Belief* (Chapel Hill: University of North Carolina Press, 1955), 20. As Coulson noted, "Gaps of this sort have the unpre-ventable habit of shrinking." Fuller argued against scientists' objections to teaching creationism in public schools as early as 1996. See Steve Fuller, "Does Science Put an End to History, Or History to Science?" in *Science Wars*, ed.

Andrew Ross (Durham, N.C.: Duke University Press, 1996), 49.

14.   C. S. Lewis, *The Great Divorce* (New York: Harper and Row, 1946), ix.

15.   Gould, *Hen's Teeth and Horse's Toes*, 275.

16.   Lucretius, *On the Nature of the Universe* (verse translation) (Oxford: Oxford University Press, 1999), 7–8 (1.145–158).

17.   David C. Lindberg, *The Beginnings of Western Science*, 2nd ed. (Chicago: University of Chicago Press, 2007), 364–67.

18.   The conception of "matter" has always been a dynamic one that has advanced with science. On this see Richard C. Vitzthum, *Materialism: An Affirmative History and Definition* (Amherst, N.Y.: Prometheus Books, 1995), 176–80. On scientific realism see Jean Bricmont and Alan Sokal, "Defense of a Modest Scientific Realism," in Sokal, *Beyond the Hoax*, 229–58.

19.   Bertrand Russell, "Introduction," in Frederick Albert Lange, *The History of Materialism* (New York: Humanities Press, 1950), v. For more complete discussions of materialism see John Bellamy Foster, *Marx's Ecology: Materialism and Nature* (New York: Monthly Review Press, 2000), 2–9; Roy Bhaskar, "Materialism," in *A Dictionary of Marxist Thought*, ed. Tom Bottomore (Oxford; Blackwell, 1983), 324.

20.   Frederick Engels, *Ludwig Feuerbach and the Outcome of Classical German Philosophy* (New York: International Publishers, 1941), 21.

21.   Richard Levins and Richard Lewontin, *The Dialectical Biologist* (Cambridge, Mass.: Harvard University Press, 1985), 272–74; Richard Lewontin and Richard Levins, *Biology Under the Influence* (New York: Monthly Review Press, 2007), 101–24. See also Brett Clark and Richard York, "Dialectical Materialism and Nature," *Organization & Environment* 18/3 (Spring 2005): 318–37; Ardea Skybreak, *The Science of Evolution and the Myth of Creationism* (Chicago: Insight Press, 2006), 299–300.

22.   See below chapter 3.

23.   Robert T. Pennock, *Tower of Babel: The Evidence Against the New Creationism* (Cambridge, Mass.: MIT Press, 1999), 194–200; Vitzthum, *Materialism*, 15, 179; National Academy of Sciences, *Science, Evolution and Creationism* (Washington, D.C.: National Academies Press, 2008).

24.   Lewontin and Levins, *Biology Under the Influence*, 92.

25.   Gould, *Rocks of Ages*, 3–6, 65–66, 84; Gould, *The Hedgehog, the Fox, and the Magister's Pox* (New York: Harmony Books, 2003), 142-143, 242-243.

26.   In *The Abolition of Man*, C. S. Lewis attempted to argue that there was a universal anthropology of morals that gave credence to the Christian Logos (morality of morals). His analysis, however, has to be judged as a failure, given the vast cultural diversity of human societies over time. That is, he failed to derive the *ought* of Christianity from the *is* of human society. See Lewis, *The Abolition of Man* (New York: Harper and Row, 1947).

27.   The emphasis on the social-contractural, not foundationalist, basis of morality goes back to Epicurus, who introduced the concept of "social contract." See

Epicurus, *The Epicurus Reader* (Indianapolis: Hackett Publishing, 1994), 32–36. It later influenced both British utilitarianism and Marx's radical historicist approach to ethics. See esp. Cornel West, *The Ethical Dimensions of Marxist Thought* (New York: Monthly Review Press, 1991).

28.   Stephen Jay Gould, "Foreword: A Life's Epistolary Drama," in *Origins: Selected Letters of Charles Darwin, 1822–1859*, ed. Frederick Burkhardt (Cambridge: Cambridge University Press, 2008), ix–xxi; Sokal, *Beyond the Hoax*, 347. Sokal was replying directly to Gould's NOMA.

29.   Rachel Carson, *Lost Woods* (Boston: Beacon Press, 1998), 210.

30.   See Eugenie C. Scott, *Evolution vs. Creationism* (Berkeley: University of California Press, 2004), 252–53.

31.   Phillip E. Johnson, *The Wedge of Truth: Splitting the Foundations of Naturalism* (Downers Grove, Ill.: InterVarsity Press, 2000), 99–101.

32.   Quoted in Janet Browne, *Charles Darwin: The Power of Place* (New York: Knopf, 2002), 106. Of course, Huxley was not above ducking the problem of materialism and science with his concept of agnosticism.

33.   William A. Dembski, "Foreword," in Benjamin Wicker, *Moral Darwinism* (Downers Grove, Ill.: InterVarsity Press, 2002), 10.

34.   Benjamin Wiker, *Moral Darwinism: How We Became Hedonists* (Downers Grove, Ill.: InterVarsity Press, 2002), 24; also see Benjamin Wiker and Jonathan Witt, *A Meaningful World* (Downers Grove, Ill.: InterVarsity Press, 2006), 15–16.

35.   Benjamin Wiker, "Darwin as Epicurean: Interview," *Touchstone* 15/8 (October 2002), http://www.touchstone.mag/. In this book we follow intelligent design proponents in applying the term "intelligent design" not simply to arguments regarding the natural and physical world but also to the notion that the social and cultural world is intelligently designed.

36.   Stephen Jay Gould, *Ever Since Darwin* (New York: W. W. Norton, 1977), 13.

37.   Cyril Bailey, "Karl Marx on Greek Atomism," *Classical Quarterly* 22/3–4 (1928): 205–6; Benjamin Farrington, *The Faith of Epicurus* (New York: Basic Books, 1967); Paul M. Schafer, "The Young Marx on Epicurus," in *Epicurus: His Continuing Influence and Contemporary Relevance*, ed. Dane R. Gordon and David B. Suits (Rochester, N.Y.: Rochester Institute of Technology, Cary Graphic Arts Press, 2003), 127–38; Foster, *Marx's Ecology*, 21–65.

38.   Epicurus in *The Hellenistic Philosophers*, ed. A. A. Long and David Sedley (Cambridge: Cambridge University Press, 1987), 1:102.

CHAPTER 2: THE WEDGE STRATEGY

1.   Eugenie Scott, *Evolution vs. Creationism* (Berkeley: University of California Press, 2004), 128.

2.   Phillip E. Johnson, *The Wedge of Truth: Splitting the Foundations of Naturalism* (Downers Grove, Ill.: InterVarsity Press, 2000), 93.

3.   Phillip E. Johnson, "Foreword," in William A. Dembski and Jay Wesley

Richards, *Unapologetic Apologetics* (Downers Grove, Ill.: InterVarsity Press, 2001), 9; William A. Dembski, ed., *Mere Creation: Science, Faith, and Intelligent Design* (Downers Grove, Ill.: InterVarsity Press, 1998); Barbara Forrest and Glenn Branch, "Wedging Creationism into the Academy," Academe Online (2005), http://www.aaup.org/AAUP/pubsres/acad eme/2005/JF/Feat/ forr.htm?PF=1.

4. Pew Research Center Pollwatch, "Reading the Polls on Creation and Evolution," September 28, 2005, http://people-press.org; Ronald L. Numbers, *The Creationists: From Scientific Creationism to Intelligent Design* (Cambridge, Mass.: Harvard University Press, 2006), 1.

5. Constance Holden, "Republican Candidate Picks Fight with Darwin," *Science* 209/4462 (September 12, 1980): 1214.

6. On the funding of the Discovery Institute see Jodi Wilgoren, "Politicized Scholars Put Evolution on the Defensive," *New York Times*, August 21, 2005.

7. Francis X. Clines, "Capitol Sketchbook; In a Bitter Cultural War, An Ardent Call to Arms," *New York Times*, June 17, 1999; Denyse O'Leary, *By Design or By Chance?* (Minneapolis: Augsburg Books, 2004), 161; William F. Buckley, "So Help Us Darwin," *National Review Online*, February 16, 2007; Ronald Bailey, "Evolutionary Politics" (January 8, 2008), reasononline, http://wwww.reason.com; Forrest and Branch, "Wedging Creationism into the Academy." For Gilder's views on intelligent design see George Gilder, "The Materialist Superstition," *Intercollegiate Review* 31/2 (Spring 1996): 6–14.

8. Johnson has sometimes referred to the thin end of the wedge as his attack on the metaphysical naturalist roots of modern science, after which the thickening of the wedge takes the form of first the presentation of intelligent design as "science," and then, as the wedge is hammered in fully, a general splitting of the "log" of materialist philosophy, where it is most important: theology, culture, and social science. See Johnson, *The Wedge of Truth*, 13–16.

9. William A. Dembski, *Intelligent Design: The Bridge Between Science and Theology* (Downers Grove, Ill.: InterVarsity Press, 1999), 120.

10. Johnson, *The Wedge of Truth*, 14.

11. Center for Renewal of Science and Culture, "The Wedge Strategy," February 5, 1999 (date first posted on the Web), http://www.antievolution.org/features/ wedge.html.

12. Center for Renewal of Science and Culture, "The Wedge Strategy"; Johnson, *The Wedge of Truth*, 13.

13. Johnson, *The Wedge of Truth*, 16, 152, 167; Forrest and Branch, "Wedging Creationism into the Academy."

14. Wilgoren, "Politicized Scholars." Reverend Moon "assigned Johnathan Wells the task of destroying evolution." And Wells noted that he became convinced that he had to devote his "life to destroying Darwinism, just as many of my fellow Unificationists had already devoted their lives to destroying Marxism." See Michael Shermer, *Why Darwin Matters* (New York: Henry Holt, 2006), 110.

15. John G. West, "The Death of Materialism and the Renewal of Culture: Introduction," *Intercollegiate Review* 31/2 (Spring 1996): 3–5; Barbara Forrest and Paul R. Gross, *Creationism's Trojan Horse: The Wedge of Intelligent Design* (Oxford: Oxford University Press, 2007), 31.

16. Bruce Chapman, "Postscript," in *Mere Creation*, ed. William A. Dembski (Downers Grove, Ill.: InterVarsity Press, 1998), 457–58.

17. Roger Downey, "Discovery's Creation," *Seattle Weekly*, February 1, 2006, http://www.seattleweekly.com.

18. For details including the quote from Richards taken from a letter by James Still, former editor of the Secular Web, see Forrest and Gross, *Creationism's Trojan Horse*, 25–33; see also Wilgoren, "Politicized Scholars."

19. Scott, *Evolutionism vs. Creationism*, 124–25.

20. John G. West, "C. S. Lewis and the Materialist Menace," July 15, 1996, address delivered at Seattle Pacific University, http://www.discovery.org; "Finding the Permanent in the Political: C. S. Lewis as a Political Thinker," http://www.discovery.org/a/457; and "The Death of Materialism and Renewal of Culture: Introduction"; Phillip Johnson, "C. S. Lewis, *That Hideous Strength (1945)*," *First Things* (March 2000), 48–49, http://www.leaderu.com/ftissues/ft0003/articles/lewis.html; C. S. Lewis, *That Hideous Strength* (London: Macmillan, 1946), 70.

21. Jay Wesley Richards, "Naturalism in Theology and Biblical Studies," in *Unapologetic Apologetics: Meeting the Challenge of Theological Studies*, ed. William A. Dembski and Jay Wesley Richards (Downers Grove, Ill.: InterVarsity Press, 2001), 97; C. S. Lewis, *Miracles* (New York: Harper and Row, 1960), 6.

22. Richards, "Naturalism in Theology and Biblical Studies," 97–98, 103; William A. Dembski, "The Problem of Error in Scripture," in Dembski and Richards, eds., *Unapologetic Apologetics*, 94; West, "Finding the Permanent in the Political," http://www.discovery.org/a/457; Lewis, *God in the Dock* (Grand Rapids, Mich.: Eerdmans, 1970), 91–92.

23. C. S. Lewis, *The Seeing Eye* (New York: Ballantine Books, 1967), 80; and *Surprised by Joy* (New York: Harcourt, 1955), 139–44.

24. Lewis, *Miracles*, 5–28; C. S. Lewis, *God in the Dock* (Grand Rapids: Eerdmans, 1970), 208–11.

25. Lewis, *Miracles*, 267–70, 13, 49, 102; Lewis, *God in the Dock*, 208–11.

26. Lewis, *Miracles*, 39, 48–49, 205.

27. Ibid., 169, 173, 279. The anti-materialist and anti-scientific aspects of C. S. Lewis's thought raised the ire of J. B. S. Haldane, one of the great Darwinian thinkers of the twentieth century (a key contributor to the development of the neo-Darwinian synthesis), who directly criticized Lewis's views. See J. B. S. Haldane, "Auld Hornie, F.R.S.," *Modern Quarterly*, no. 4 (Autumn 1946), 32–40.

28. Nancy Pearcey, *Total Truth: Liberating Christianity from its Cultural Captivity* (Wheaton, Ill: Good News Publications, 2004), 389–92.

29.   In a sense, it might be argued that the whole debate extends back further to the early atomists, Leucippus and Democritus, who were quite clearly challenging design notions, and to which Socrates was undoubtedly responding. Thus both sides of the argument can be traced back to ancient Greek philosophers of the fifth century BCE. See David Sedley, *Creationism and Its Critics in Antiquity* (Berkeley: University of California Press, 2007), 133–35.

CHAPTER 3: EPICURUS'S SWERVE

1.   David Hume, *Enquiries Concerning Human Understanding and Concerning the Principles of Morals* (Oxford: Oxford University Press, 1975), 135–140. For a further discussion of Hume's position see chapter 4 in this volume.

2.   Charles Darwin, *The Correspondence of Charles Darwin* (Cambridge: Cambridge University Press, 1994), 9:135. For a more detailed treatment of Darwin's critique of the concept of "intelligent design" see chapter 6.

3.   See the timeline at http://www.researchintelligentdesign.org/wiki/Intelligent _Design_Timeline.

4.   Spencer R. Weart, *The Discovery of Global Warming* (Cambridge, Mass.: Harvard University Press, 2003), 3–4.

5.   Lucretius is one of our main sources of Epicurus's ideas, whose own writings have been largely lost. Lucretius's epic poem has been viewed by classical scholars over the centuries as a faithful attempt accurately to convey in verse form Epicurus's natural philosophy, specifically the latter's *On Nature*. We follow the normal practice of classical scholars therefore in attributing the specific ideas and statements to be found in Lucretius to Epicurus.

6.   John Tyndall, *Fragments of Science*, vol. 2 (New York: D. Appleton and Co., 1896), 143 (emphasis added). Marx and Engels were familiar with Tyndall's Belfast Address and supported its main propositions, though they were critical of Tyndall for not being materialist enough. Tyndall's argument on Epicurus was heavily based on Frederick Lange, *The History of Materialism* (New York: Humanities Press, 1950), first published in German in 1865. Tyndall's Belfast Address has been targeted by intelligent design proponents. Thus William Dembski attacks Tyndall for declaring, "Science demands the . . . absolute reliance upon laws in nature." See William A. Dembski, *Intelligent Design: The Bridge Between Science and Theology* (Downers Grove, Ill.: InterVarsity Press, 1999), 89.

7.   David Sedley, *Creationism and Its Critics in Antiquity* (Berkeley: University of California Press, 2007), 78. It has been common to attribute the origin of the argument from design to Socrates's contemporary, Diogenes of Appolonia (mid-fifth century BCE), based on a few fragments. Diogenes was influenced by Leucippus's work and was undoubtedly replying like Socrates himself to the atomists. Sedley argues convincingly, however, that the fragments from Diogenes are somewhat ambiguous, while Socrates generated a clearer, more complete and much more influential argument from design at about the same

time and thus must be seen as its originator (Ibid., 75–78, 90).

8.   Xenophon, *Conversations of Socrates* (London: Penguin Books, 1990), 90–91 (*Memorabilia*, 1.4.3–1.4.11).

9.   Xenophon, *Conversations of Socrates*, 90–92, 192–93 (*Memorabilia*, 1.4.3–1.4.11, 4.3.6–4.3.13).

10.   Sedley, *Creationism and Its Critics in Antiquity*, 82–83, 86.

11.   Plato, *Timaeus and Critias* (London: Penguin Books, 1970), 42, 54, 96; Plato, *The Laws* (London: Penguin Books, 1970), 415–20; Sedley, *Creationism and Its Critics in Antiquity*, 98–109; Benjamin Farrington, *Science and Politics in the Ancient World* (London: George Allen and Unwin, 1939), 131–32; Benjamin Farrington, *The Faith of Epicurus* (London: Weidenfeld and Nicolson, 1967), 73.

12.   Aristotle, *The Basic Works of Aristotle* (New York: Random House, 1941), 644–45 (1.1. 639–50); Sedley, *Creationism and Its Critics in Antiquity*, 167–73.

13.   Sedley, *Creationism and Its Critics in Antiquity*, 210–25.

14.   Cicero, *The Nature of the Gods* (London: Penguin, 1972), 159–63 (II.87–98).

15.   Cicero belonged to the philosophical school known as the Academy. He was a political opponent in the Roman Senate of the wealthy Lucius Calpurnius Piso Caesoninus (first century BCE), the father-in-law of Julius Caesar, and a leading Epicurean of ancient Rome. Cicero's anti-Epicureanism is thus commonly believed to have been targeted at Piso (and through him at Caesar). Lucius Calpurnius Piso is believed to have been the owner of the Villa of the Papyri in Herculaneum, where an Epicurean library of some 1,200 books, collected by the Greek Epicurean philosopher Philodemus (c. 110–37 BCE) under Piso's sponsorship was established, and which was destroyed with the eruption of Vesuvius in 79 CE. The charred remains of the papyri, discovered in the eighteenth century, many of which have been gradually unrolled and partially deciphered by laborious scientific techniques since then, represent the only surviving library from the ancient world. A new technology of multi-spectral imaging now makes it possible to read the papyri without unrolling and damaging them. See John T. Fitzgerald, "Introduction," in *Philodemus in the New Testament World*, ed. Fitzgerald, Dirk Obbink, and Glenn S. Holland (Boston: Brill, 2004), 6–7.

16.   A. A. Long, "Evolution vs. Intelligent Design in Classical Antiquity," http://townsendcenter.berkeley.edu/article2.shtml (November 2006); and *From Epicurus to Epictetus* (New York: Oxford University Press, 2006), 155–77.

17.   See Elizabeth Asmis, *Epicurus's Scientific Method* (Ithaca, N.Y.: Cornell University Press, 1984); Philodemus, *On Methods of Inference: A Study of Ancient Empiricism* (Philadelphia: American Philological Association, 1941). Epicurus's emphasis on empirical knowledge derived from the senses and de-emphasis on mathematics, which in his day was used mainly to promote astral

religion, led him to a grotesque conclusion on the size of the sun, saying that it was the size that it seemed to be. He was undoubtedly drawn to this by his determined opposition to Plato's proposed state religion, which made gods of the sun and moon. See Farrington, *Science and Politics*, 139–41.

18. Sedley, *Creationism and Its Critics in Antiquity*, 141.

19. Lucretius, *On the Nature of the Universe* (London: Penguin, 1994), 13 (1.149).

20. George K. Strodach, "Introduction," in Epicurus, *The Philosophy of Epicurus* (Easton, Penn.: Northwestern University Press, 1963), 45–50, 234–35; Epicurus, *The Epicurus Reader* (Indianapolis: Hackett Publishing, 1994), 19–28, 34; Asmis, *Epicurus's Scientific Method*, 321–30.

21. Lucretius, *On the Nature of the Universe* (Penguin), 22 (1.475–82).

22. Arend Th. Van Leeuwen, *Critique of Heaven* (New York: Charles Scribner's Sons, 1972), 102–07.

23. J. M. Rist, *Epicurus: An Introduction* (Cambridge: Cambridge University Press, 1972), 10–11; Pamela Gordon, "Remembering the Garden: The Trouble with Women in the School of Epicurus," in Fitzgerald et al., *Philodemus and the New Testament World*, 221–43; Farrington, *The Faith of Epicurus*, 30; A. J. Festugière, *Epicurus and His Gods* (New York: Russell & Russell, 1955), 29–31.

24. Jean–Paul Sartre, *Literary and Philosophical Essays* (New York: Criterion Boos, 1955), 207.

25. Karl Marx and Frederick Engels, *Collected Works* (New York: International Publishers, 1975), 1:51. Marx's comparison to "the plastic gods of Greek art" captured the essence of the Epicurean view, since Epicurus argued that the gods existed in human form and through preconceptions in their dreams human beings were introduced to divine nature and thus to moral (and aesthetic) standards for which to strive. In the twentieth century, knowledge of Epicurean theology expanded enormously as a result of new materials obtained from the papyri in Herculaneum. Recently, some Epicurean scholars have argued that the gods were merely "thought constructs" for Epicurus, though this is strongly disputed by others. See A. A. Long, and D. N. Sedley, *The Hellenistic Philosophers*, vol. 1 (Cambridge: Cambridge University Press, 1987), 145–49; and James A. Flippin's remarkable B.A. thesis, "Epicurus Transformed: Epicurean Religion in Lucretius' Roman Context" (classics, Reed College, May 2006).

26. Alfred Lord Tennyson, *The Poems of Tennyson* (Berkeley: University of California Press, 1987), 713.

27. Farrington, *Faith of Epicurus*, 72–3, Festugière, *Epicurus and His Gods*, 63–65, 73–89; Epicurus, *The Epicurus Reader*, 96; Lucretius, *On the Nature of the Universe* (Oxford: Oxford University Press, 1999), 169–72 (5.1161–1240); Philodemus, *On Piety, Part 1*, ed. with commentary by Dirk Obbink (Oxford: Oxford University Press, 1996), 9–14, 121, 157–59, 282–83, 306. Epicurus claimed that the knowledge of the gods came simply through preconceptions

(*prolepses*) provided in dreams in which the gods took anthropomorphic forms. The most uncluttered preconceptions of the gods seem to have been provided in the dreams of the early humans. The preconceptions of the gods provided insights into divine life and character and moral ideals for human action, and the gods themselves became objects of worship, but they did not otherwise impinge on the material world.

28. Stephen Jay Gould, *Rocks of Ages* (New York: Ballantine, 1999).

29. Farrington, *Faith of Epicurus*, 64–65.

30. Howard Jones, *The Epicurean Tradition* (New York: Routledge, 1992), 97.

31. St. Augustine, *City of God* (London: Penguin, 1984), 309 (VIII.7.1), 432 (XI.4.2); Thomas Aquinas, *An Aquinas Reader* (Garden City, N.Y.: Doubleday, 1972), 161, 336–37.

32. Dante, *The Inferno* (New York: Signet, 1982), 96.

33. Plato, *Laws*, 437.

34. See A. A. Long, *Hellenistic Philosophy: Stoics, Epicureans, Sceptics* (Berkeley: University of California Press, 1986), 108, 144–52.

35. Norman W. DeWitt, *Epicurus and His Philosophy* (Minneapolis: University of Minnesota Press, 1954), 275.

36. Georg Willhelm Friedrich Hegel, *Lectures on the History of Philosophy* (Lincoln: University of Nebraska Press, 1965), 2:297; John Bellamy Foster, *Marx's Ecology* (New York: Monthly Review Press), 2–9, 39–51.

37. William A. Dembski, "Foreword," in Benjamin Wiker, *Moral Darwinism: How We Became Hedonists* (Downers Grove, Ill.: InterVarsity Press, 2002), 10–11.

38. Phillip E. Johnson, *The Wedge of Truth: Splitting the Foundations of Naturalism* (Downers Grove, Ill.: InterVarsity Press, 2000), 99–100.

39. Wiker, *Moral Darwinism*, 27, 149, 314–15; Phillip E. Johnson, "The Church of Darwin," *Wall Street Journal*, August 16, 1999.

40. Marx and Engels, *Collected Works*, 1:70–72.

41. Epicurus, *The Epicurus Reader*, 32.

42. Lucretius, *On the Nature of the Universe* (Penguin), 149–50 (5.830–831).

43. Thomas S. Hall, *Ideas of Life and Matter: Studies in the History of General Physiology, 1600 B.C.–1900 A.D.*, vol. 1 (Chicago: University of Chicago Press, 1969), 19–20.

44. William A. Dembski, *No Free Lunch* (New York: Rowman and Littlefield, 2002), 1.

45. Lucretius, *On the Nature of the Universe* (Penguin), 149 (5.790–825).

46. For the history and development of scientific views on the origin of life, attempting to explain life's emergence, see J. D. Bernal, *The Origins of Life* (London: Weidenfeld and Nicolson, 1967); Robert M. Hazen, *Genesis: The Scientific Quest for Life's Origin* (Washington, D.C.: Joseph Henry Press, 2005); and Antonio. Lazcano, "Creationism and the Origin of Life," in *Scientists Confront Intelligent Design and Creationism*, ed. A. J. Petto and L. R. Godfrey (New York: W. W. Norton, 2007), 180–96. It is no accident that intel-

ligent design proponents have attacked the Haldane-Oparin theory, and later scientific approaches it helped inspire, simply dismissing it as derived from Epicurus, Darwin, Engels, and Marx. See Benjamin Wiker and Jonathan Witt, *A Meaningful World* (Downers Grove, Ill.: InterVarsity Press, 2006), 199–219.

47. John D. Barrow and Frank J. Tipler, *The Anthropic Cosmological Principle* (Oxford: Oxford University Press, 1986), 34; Aristotle, *Basic Works*, 249.

48. Lucretius, *On the Nature of the Universe* (Penguin), 150–51 (5.836–876); Michael Ruse, *Darwin and Design* (Cambridge, Mass.: Harvard University Press, 1999), 25–26.

49. Sedley, *Creationism and Its Critics in Antiquity*, 53; Cicero, *The Nature of the Gods*, 181 (II.141–144).

50. Lucretius, *On the Nature of the Universe* (Penguin), 116–17 (5.820–850).

51. Gordon Campbell, *Lucretius on Creation and Evolution* (Oxford: Oxford University Press, 2003), 8, 261–72; Lucretius, *On the Nature of the Universe*, 154–57 (5.1012–1118). Campbell insists that Lucretius points in the case of humans to evolution within species, but nonetheless insists on the fixity of species, i.e., one species cannot transmute into another.

52. Sedley, *Creationism and Its Critics in Antiquity*, 155–66; Long and Sedley, *The Hellenistic Philosophers*, 1:37–44.

53. Campbell, *Lucretius on Creation and Evolution*, 264; Lucretius, *On the Nature of the Universe* (Penguin), 130–31 (5.1091–1105).

54. Lucretius, *On the Nature of the Universe* (Penguin) 152–66 (5.925–1456); Benjamin Farrington, *Science and Politics*, 125–26.

55. Epicurus, *The Epicurus Reader*, 36. We made a slight change in punctuation in the translation for readability.

56. Marx and Engels, *Collected Works*, 5:141–42; Farrington, *Science and Politics*, 159; Campbell, *Lucretius on Creation and Evolution*, 253–54.

57. Marx and Engels, *Collected Works*, 5:141–42; Norman W. DeWitt, *St. Paul and Epicurus* (Minneapolis: University of Minnesota Press, 1954).

58. Marx and Engels, *Collected Works*, 1: 30, 49–53, 73, 5:141–52; Epicurus, *The Epicurus Reader*, 29.

CHAPTER 4: ENLIGHTENMENT MATERIALISM
AND NATURAL THEOLOGY

1. Peter Gay, *The Enlightenment: An Interpretation* (New York: Alfred A. Knopf, 1966), 1:98–107, 356. Voltaire himself was a deist, strongly influenced by Newton, and critical of outright materialism.

2. Howard Jones, *The Epicurean Tradition* (New York: Routledge, 1989), 142–43, 164–65; Thomas Kuhn, *The Copernican Revolution* (Cambridge, Mass.: Harvard University Press, 1985), 199, 235–37; John Hedley Brooke, *Science and Religion* (New York: Cambridge University Press, 1991), 74–75; Giordano Bruno, *The Ash Wednesday Supper* (Toronto: University of Toronto Press, 1995), 206; Frederick Albert Lange, *The History of Materialism* (New

York: Humanities Press, 1950), 23–35.

3.   Francis Bacon, *Philosophical Works* (New York: Freeport, 1905), 473.

4.   Karl Marx and Frederick Engels, *Collected Works* (New York: International
       Publishers, 1975), 4:125–26; Neal Wood, *"Tabula Rasa,* Social
       Environmentalism, and the 'English Paradigm,'" *Journal of the History of Ideas*
       53:647–68.

5.   Robert Boyle, *Works* (London: A. Millar, 1744), 4:515; Monte Johnson and
       Catherine Wilson, "Lucretius and the History of Science," in *The Cambridge
       Companion to Lucretius*, ed. Stuart Gillespie and Philip Hardie (Cambridge:
       Cambridge University Press, 2007), 139.

6.   Isaac Newton, *Newton's Philosophy of Nature* (Mineola, N.Y.: Dover, 1953),
       41–50, 177; I. Bernard Cohen and George E. Smith, "Introduction," in *The
       Cambridge Companion to Newton*, ed. Cohen and Smith (Cambridge:
       Cambridge University Press, 2002), 23; Johnson and Wilson, "Lucretius and
       the History of Science," 141. Newton reinforced his arguments with natural-
       theological views in the biological realm, pointing to "the uniformity in the
       bodies of animals" as evidence of design. But this lay outside of his principal
       area of scientific investigation. Ibid., 65, 177–78.

7.   Gottfried Wilhelm Leibniz and Samuel Clarke, *The Leibniz–Clarke
       Correspondence, Together with Extracts from Newton's* Principia *and* Optiks
       (Manchester: Manchester University Press, 1956); Domenico Bertolini Meli,
       "Newton and the Leibniz–Clark Correspondence," in *The Cambridge
       Companion to Newton*, 455–64.

8.   Gottfried Wilhelm Leibniz, *Discourse on Metaphysics and Other Essays*
       (Indianapolis: Hackett, 1991), 55. Leibniz in the italicized passage is quoting
       from Virgil, *The Eclogues and the Georgics* (Cambridge: Cambridge University
       Press, 1944), 312 (*Georgics* IV, 393).

9.   H. G. Alexander, "Introduction," in Leibniz and Clarke, *Leibniz–Clarke
       Correspondence*, xvii.

10.   Leibniz and Clarke, *Leibniz–Clarke Correspondence*, 11–12. Leibniz's criticism
       of Newton's conception of a "mending" God is strongly supported by intelli-
       gent design proponents Nancy Pearcey and Charles B. Thaxton in their *The
       Soul of Science* (Wheaton, Ill.: Crossway Books, 1994), 91–92.

11.   Leibniz and Clarke, *Leibniz–Clarke Correspondence*, 11–12, 14, 39, 43–45, 96,
       172–83. Leibniz made it clear he did not deny that God intervened in the
       world, only that it was necessary for him to do so in order to "mend" his cre-
       ations. But his philosophy of pre-established harmony nonetheless gave rise to
       the notion of a perfect clockwork subordinated to final causes thereby making
       God's interventions seemingly redundant. See Frederick Copleston, *A History
       of Philosophy* (Garden City, N.Y.: Doubleday, 1963), 4:313–14.

12.   Newton, *Newton's Philosophy of Nature*, 41–46.

13.   Richard Lewontin, "Billions and Billions of Demons," *New York Review of
       Books* 64/1 (January 9, 1997): 28–32. There are a number of places in his intel-

lectual corpus in which Newton employs design inferences with regard to nature, but these tend to be brief speculative asides rather than systematic developed arguments. They appear chiefly in his correspondence, and are largely disconnected from his main scientific contributions and the *Principia*. See Newton, *Newton's Philosophy of Nature*, 58–67. It is clear that Newton, however devout, sought as a physicist to avoid compromising his natural science with supernatural explanations, and turned to the latter only as a last resort. Claims that he saw his science and theology as easily compatible and easily intermixed therefore fail to account for the full complexity of his thought.

14.   Newton quoted in Richard S.Westfall, "Newton and Christianity," in *Newton: A Norton Critical Edition*, ed. I. Bernard Cohen and Richard Westfall (New York: W. W. Norton, 1995), 357.

15.   Westfall, "Newton and Christianity," 361–63, 370; Scott Mandelbrote, "Newton and Eighteenth-Century Christianity," in Cohen and Smith, eds., *The Cambridge Companion to Newton*, 412–414.

16.   Steve Fuller, *Science vs. Religion?* (Cambridge: Polity Press, 2007), 34–36, 54–56, 100–2, 163. See also Steve William Fuller "Rebuttal of Dover Expert Reports" (Expert Testimony for Defense), *Kitzmiller et al. v. Dover Area School District* (submitted May 13, 2005), 7–8; and Steve Fuller, "Schools for the Enlightenment or Epiphany," *The Times Higher Education Supplement*, November 19, 2007. Fuller's argument with respect to Newton and intelligent design consists of repeatedly asserting (devoid of all evidence) that Newton "arrogantly" believed he had penetrated "the mind of God" and understood his design. Whatever this might mean, it obviously has no bearing on the logical bases of Newton's scientific discoveries. That religion was an external sociological factor in Newton's day affecting the development of science no one would doubt, and was long ago established by Robert Merton in his *Social Theory and Social Structure* (Glencoe, Ill.: Free Press, 1957). That religion or the Bible constituted a "sure path" to science as Fuller contends is quite another matter.

17.   Pearcey and Thaxton, *The Soul of Science*, 89–91.

18.   C. A. Coulson, *Science and Christian Belief* (Chapel Hill: University of North Carolina Press, 1955), 20–21.

19.   Stephen Jay Gould, *Dinosaur in a Haystack* (New York: Harmony Books, 1995), 24–37. Ironically, Pearcey and Thaxton as defenders of Leibniz versus Newton argue that "Laplace was right to reject this particular notion of divine activity [God's role in repairing the celestial orbits] that Newton had advocated. God does not act to repair leaks and cracks in His original creation." See Pearcey and Thaxton, *The Soul of Science*, 92.

20.   John Locke, *The Reasonableness of Christianity as Delivered in the Scriptures* (Oxford: Oxford University Press, 1999), 143.

21.   John Ray, *The Wisdom of God Manifested in the Works of Creation* (London: Benjamin Walford, 1699), 35–39, 41, 49, 53, 81, 116, 257, 425; John C. Greene, *The Death of Adam: Evolution and its Impact on Western Thought*

(Ames, Iowa: Iowa State University Press, 1959), 1–10.

22.   William Paley, *Natural Theology—or Evidence of the Existence and Attributes of the Deity Collected from the Appearances of Nature* (London: R. Faulder, 1803), 9, 473. See also Stephen Jay Gould, *Eight Little Piggies* (New York: W. W. Norton, 1993), 150–51, 139–40.

23.   William Paley, *The Principles of Moral and Political Philosophy* (New York: Harper and Brothers, 1867), 36–38, 44, 99–103, 278.

24.   Paley, *Natural Theology*, 539–42. Intelligent design proponent William Dembski acknowledges that the natural theology tradition was the forerunner of today's intelligent design argument. But natural theology of the sort promoted by Paley, he points out, failed largely because it produced one "false positive" after another, with its purported evidence of God's contrivances in nature being shown again and again to be subject to naturalistic explanation. Today's intelligent design inference, with its emphasis on specified irreducible complexity, we are told, does not have the same weaknesses—nor is it as directly theological. See William A. Dembski, *Intelligent Design: The Bridge Between Science and Theology* (Downers Grove, Ill.: InterVarsity Press, 1999), 70–93.

25.   Thomas Malthus, *An Essay on the Principle of Population and a Summary View of the Principle of Population* (Harmondsworth: Penguin, 1970), 201–12, 272; Thomas Malthus, *An Essay on the Principle of Population; or A View of its Past and Present Effects on Human Happiness; With an Inquiry into Our Prospects Respecting the Future Removal or Mitigation of the Evils which it Occasions* (Cambridge: Cambridge University Press, 1989), 2:140–41, 101–5.

26.   Malthus, *An Essay on the Principle of Population and a Summary View*, 204–5.

27.   John Bellamy Foster, *Marx's Ecology: Materialism and Nature* (New York: Monthly Review Press, 2000), 86–102.

28.   Thomas Malthus, *The Unpublished Papers in the Collection of Kanto Gakuen University* (Cambridge: Cambridge University Press, 2004), 1–24. Malthus's argument that natural theology and scripture/revelation all pointed to the same conclusions regarding God and population can also be seen in his argument in *A Summary View*: *An Essay on the Principle of Population and a Summary View*, 271–72.

29.   James Secord, *Victorian Sensation* (Chicago: University of Chicago Press, 2000), 276–77.

30.   Stephen Jay Gould, *The Structure of Evolutionary Theory* (Cambridge, Mass.: Harvard University Press, 2002), 117.

31.   Secord, *Victorian Sensation*, 279.

32.   Thomas Chalmers, *On the Power, Wisdom and Goodness of God as Manifested in the Adaptation of External Nature to the Moral and Intellectual Constitution of Man* (London: William Pickering, 1834), 1:17–21.

33.   Secord, *Victorian Sensation*, 273.

34.   Chalmers, *On the Power, Wisdom and Goodness of God*, 1:22, 252; 2:7, 34–35; Chalmers, *On Political Economy in Connexion with the Moral State and Moral*

Prospects of Society (Glasgow: William Collins, 1853), 2:338.

35. David Hume, *Enquiries Concerning Human Understanding and Concerning the Principles of Morals* (Oxford: Oxford University Press, 1975), 132–48. Hume is not concerned primarily with accuracy with regard to Epicurus's views in this speech, whom he is using in this case to expound his own views. Thus Hume is doubtless aware that Epicurus did not altogether deny the existence of the gods.

36. David Hume, *Dialogues Concerning Natural Religion* (Indianapolis: Bobbs–Merrill, 1947), 176–87. Dembski refers to this set of arguments as Hume's "Epicurean move." See Dembski, *Intelligent Design*, 286.

37. James Boswell, "An Account of My Last Interview with David Hume, ESQ, March 3, 1777," in Hume, *Dialogues Concerning Natural Religion*, 76–79; Adam Smith, "Letter from Adam Smith, LL.D., to William Strahn, Esq., November 9, 1776," in Hume, *Dialogues*, 243–48; Gay, *Enlightenment*, 1: 356.

38. Voltaire, *Candide* (London: Penguin, 1947), 20, 136.

CHAPTER 5: MARX'S CRITIQUE OF HEAVEN
AND CRITIQUE OF EARTH

1. Karl Marx and Frederick Engels, *Collected Works* (New York: International Publishers, 1975), 1:190, 493–96. Spiritual reason, Marx went on to observe here, was "classically expressed" by the early Church father, Tertullian, who claimed "it is true because it is absurd."

2. Marx and Engels, *Collected Works*, 1:73.

3. Center for Renewal of Science and Culture, Discovery Institute, *The Wedge Strategy [Document]*, 1999, http://www.antievolution.org/features/wedge. html; Benjamin Wiker and Jonathan Witt, *A Meaningful World* (Downers Grove, Ill.: Intervarsity Press, 2006), 15–16, 60.

4. Karl Marx, *Early Writings* (London: Penguin, 1974), 243–45.

5. David McLellan, *Karl Marx* (New York: Harper and Row, 1973), 1–16; Arend Th. van Leeuwen, *Critique of Heaven* (New York: Charles Scribner's Sons, 1972), 40–43, 78; Marx and Engels, *Collected Works*, 1:636–39; Sidney Hook, *From Hegel to Marx* (Ann Arbor: University of Michigan Press, 1962), 81.

6. Marx and Engels, *Collected Works*, 1:10–21.

7. Marx and Engels, *Collected Works*, 4:126–28; Arend Th. van Leeuwen, *Critique of Earth* (New York: Charles Scribner's Sons, 1974), 14–16.

8. Francis Bacon, *Philosophical Works* (New York: Freeport, 1905), 473; Marx and Engels, *Collected Works*, 1:201; Charles Darwin, *Notebooks, 1837–1844* (Ithaca, N.Y.: Cornell University Press, 1987), 637.

9. Georg Wilhelm Friedrich Hegel, *Lectures on the History of Philosophy* (Lincoln: University of Nebraska Press, 1995), 3:184–87.

10. Marx and Engels, *Collected Works*, 3:419; Frederick Engels, *Ludwig Feuerbach and the Outcome of Classical German Philosophy* (New York: International Publishers, 1941), 17, 21.

11. Charles H. Talbert, "Introduction," Hermann Samuel Reimarus, *Reimarus: Fragments* (Chico, Calif.: Scholars Press, 1970), 1; Julian Jaynes and William Woodward, "In the Shadow of the Enlightenment: I. Reimarus Against the Epicureans," *Journal of the History of the Behavioral Sciences* 10/1 (January 1974): 5–6.

12. The full German title of Reimarus's 1760 *Drives of Animals* was: *Allgemeine Betrachtungen über die Triebe der Thiere, hauptsächlich über ihre Kunsttriebe, zur Erkenntnis des Zusammenhanges zwischen dem Schöpfer und uns selbst.*

13. Hermann Samuel Reimarus, *The Principal Truths of Natural Religion Defended and Illustrated, in Nine Dissertations: Wherein the Objections of Lucretius, Buffon, Maupertuis, Rousseau, La Mettrie, and Other Ancient and Modern Followers of Epciurus are Considered, and their Doctrines Refuted* (London: B. Law, 1766), 117–120, 152, 220.

14. Ibid., 76, 84, 242.

15. Ibid., 156–57, 229–34, 250–51.

16. Jaynes and Woodward, "In the Shadow of the Enlightenment: I. Reimarus Against the Epicureans," 4; Julian Jaynes and William Woodward, "In the Shadow of the Enlightenment: II. Reimarus and His Theory of Drives," *Journal of the History of the Behavioral Sciences* 10/2 (April 1974), 154.

17. See Karl Marx, *Capital* (London: Penguin, 1976), 1:283–84; Marx, *Early Writings*, 328–29; van Leeuwen, *Critique of Earth*, 20, 53. The importance that Reimarus assumed among the Young Hegelians is evident in the work of Strauss, who wrote a major study on Reimarus's *Apology*, in which he declared "all positive religions without exception are works of deception." David Strauss, "Hermann Samuel Reimarus and His Apology," in Reimarus, *Reimarus: Fragments*, 46.

18. Marx and Engels, *Collected Works*, 4:79.

19. See Karl Marx and Friedrich Engels, *Ex Libris* (Berlin: Dietz Verlag, 1967), 127; Gottfried Wilhelm Leibniz and Samuel Clarke, *The Leibniz–Clarke Correspondence* (New York: Manchester University Press, 1956), 166; Marx and Engels, *Collected Works*, 1:190.

20. Paul M. Schafer, "Introduction," in *The First Writings of Karl Marx*, ed. Paul M. Schafer (Brooklyn, N.Y.: Ig Publishing, 2006), 45; Marx and Engels, *Collected Works*, 1:43. For a general discussion of Marx's relation to Epicurus see John Bellamy Foster, *Marx's Ecology* (New York: Monthly Review Press, 2000), 32–65.

21. Frederick Engels, "Bruno Bauer and Early Christianity," in Karl Marx and Frederick Engels, *Marx and Engels on Religion* (New York: Schocken Books, 1964), 197.

22. van Leeuwen, *Critique of Heaven*, 74.

23. Marx and Engels, *Collected Works*, 1:29–31.

24. van Leeuwen, *Critique of Heaven*, 77.

25. Marx and Engels, *Collected Works*, 1:51, 74–75, 91, 448, 508–9.

26. Marx and Engels, *Collected Works*, 1:84.

27. Marx and Engels, *Collected Works*, 1:102-5, 446–448, 509.

28. Quoted in Hook, *From Hegel to Marx*, 17.

29. Marx and Engels, *Collected Works*, 4:85; van Leeuwen, *Critique of Heaven*, 195.

30. Ernst Bloch contended that "The First and Eleventh Theses on Feuerbach are already present *in statu nascendi* in the references to Epicurus" in Marx's dissertation. See Bloch, *Karl Marx* (New York: Herder and Herder, 1971), 156.

31. The relation of Marx to Feuerbach is more complex when one considers that Feuerbach in his *History of Modern Philosophy* had dealt in detail with Gassendi's resurrection of Epicurus, and that this had influenced Marx in the writing of his dissertation. See Marx and Engels, *Collected Works*, 1:94.

32. Ludwig Feuerbach, *The Fiery Brook* (Garden City, N.Y.: Doubleday, 1972), 161, 172, 198; Foster, *Marx's Ecology*, 68-71.

33. van Leeuwen, *Critique of Heaven*, 12.

34. Marx, *Early Writings*, 243-45.

35. Marx, *Early Writings*, 357.

36. Marx, *Early Writings*, 218.

37. Thomas Dean, *Post-Theistic Thinking: The Marxist-Christian Dialogue in Radical Perspective* (Philadelphia: Temple University Press, 1975), 69.

38. Marx quoted in Eleanor Marx, "Karl Marx: A Few Stray Notes," in Institute of Marxism–Leninism, *Reminiscences of Marx and Engels* (Moscow: Foreign Languages Publishing House, n.d.), 253.

39. Marx, *Early Writings*, 421-23.

40. Marx and Engels, *Collected Works*, 5:59.

41. Karl Marx, *Capital* (London: Penguin, 1981), 3:448–49, 953–56 and 1:165.

42. Cornel West, *The Ethical Dimensions of Marxist Thought* (New York: Monthly Review Press, 1991); Marx and Engels, *Collected Works*, 6:170, 173; Epicurus, *The Epicurus Reader* (Indianapolis: Hackett Publishing, 1994), 35.

43. Marx, *Capital*, 1:766-77, 800.

44. Marx and Engels, *Collected Works*, 3:439.

45. Townsend quoted in Marx, *Grundrisse* (London: Penguin, 1973), 845.

46. Karl Marx, *Grundrisse*, 605.

47. Cobbett quoted in Kenneth Smith, *The Malthusian Controversy* (London: Routledge and Kegan Paul, 1951).

48. Karl Marx, *Theories of Surplus Value* (Moscow: Progress Publishers, 1971), pt. 1, 299-300; pt. 3, 56-57.

49. See Foster, *Marx's Ecology*, 117-26; Marx, *Collected Works*, 5:471-73.

50. Karl Liebknecht, "Reminiscences of Marx," in Institute of Marxism–Leninism, *Reminiscences of Marx and Engels*, 106; Marx and Engels, *Collected Works*, 41:232, 246-47.

51. Marx, *Capital*, 1:461, 493; Darwin, *The Origin of Species* (London: Penguin, 1968), 187-88.

52. Thomas R. Trautmann, *Lewis Henry Morgan and the Invention of Kinship*

(Berkeley: University of California Press, 1987), 35, 200; Foster, *Marx's Ecology*, 212–21.

53. Jacob W. Gruber, "Brixham Cave and the Antiquity of Man," in *Context and Meaning in Cultural Anthropology*, ed. Melford E. Spiro (New York: Free Press, 1965), 373–402. The significance of some of the early prehistoric human remains located in the nineteenth century (including those in the Neander Valley) was left in doubt at first due to the poor way in which these discoveries were excavated, deviating from the slow, careful process required by geological work, often failing to preserve the proper stratigraphic context, causing scientific observers to suspect that remains from distinct geological strata had been mingled with one another. In contrast, the excavation of the remains at Brixham Cave was supervised by the Geological Society of London and hence for the first time definitively confirmed that human beings had existed on the earth in "great antiquity."

54. Trautmann, *Lewis Henry Morgan*, 32, 172–73.

55. Marx and Engels, *Collected Works*, 5:141–42; Marx and Engels, *Collected Works*, 3:179; Foster, *Marx's Ecology*, 51–62; Frederick Engels, "Letter to Friedrich Adolph Sorge, March 15, 1883," in *Karl Marx Remembered*, ed. Philip S. Foner (San Francisco: Synthesis Publications, 1983), 26; István Mészáros, *Marx's Theory of Alienation* (London: Merlin Press, 1971), 162–89.

CHAPTER 6: ON THE ORIGIN OF DARWINISM

1. Charles Darwin, *The Autobiography of Charles Darwin, 1809–1882* (New York: W.W. Norton, 2005), 73.

2. Francis Bacon, *Philosophical Works* (New York: Freeport, 1905), 473; Charles Darwin, *Notebooks, 1836–1844* (Ithaca, N.Y.: Cornell University Press, 1987), 637.

3. John R. Durant, "The Ascent of Nature in Darwin's *Descent of Man*," in *The Darwinian Heritage*, ed. David Kohn (Princeton, N.J.: Princeton University Press, 1985), 301.

4. Adrian Desmond and James Moore, *Darwin: The Life of a Tormented Evolutionist* (New York: W. W. Norton, 1994).

5. Janet Browne, *Charles Darwin: Voyaging* (Princeton, N.J.: Princeton University Press, 1995), 72–8; Adrian Desmond, James Moore, and Janet Browne, *Charles Darwin* (Oxford: Oxford University Press, 2007).

6. Desmond and Moore, *Darwin*, 33–40; Browne, *Charles Darwin: Voyaging*, 76–8.

7. Darwin, *The Autobiography*, 50–1; Browne, *Charles Darwin: Voyaging*.

8. William Paley, *Natural Theology—or Evidence of the Existence and Attributes of the Deity Collected from the Appearances of Nature* (London: R. Faulder, 1803); Gavin de Beer, *Charles Darwin: Evolution by Natural Selection* (Garden City, N.Y.: Doubleday, 1964), 14–18.

9. Darwin, *The Autobiography*, 64, 71.

10. Ibid., 65–9; Charles Darwin, *The Voyage of the Beagle* (New York: P. F. Collier and Son, 1937).

11. Desmond and Moore, *Darwin*, 186; Darwin, *The Voyage*, 376–405; Nora Barlow, "Darwin's Ornithological Notes," *Bulletin of the British Museum (Natural History), Historical Series* 2/7 (1963): 201–78.

12. Richard York and Brett Clark, "Natural History and the Nature of History," *Monthly Review* 57/7 (2005): 21–29.

13. Frank J. Sulloway, "Darwin's Conversion: The *Beagle* Voyage and Its Aftermath," *Journal of the History of Biology* 15/3 (1982): 325–396.

14. Desmond, Moore, and Browne, *Charles Darwin*, 26–40; Browne, *Charles Darwin: Voyaging*, 362–3; Charles Darwin, *Charles Darwin's Notebooks, 1836-1844: Geology, Transmutation of Species, Metaphysical Enquiries* (Ithaca, N.Y.: Cornell University Press, 1987), Notebook B, 229.

15. Darwin, *The Autobiography*, 98.

16. Howard E. Bruber, *Darwin on Man* (Chicago: University of Chicago Press, 1981), 204–5.

17. Darwin, *Charles Darwin's Notebooks*, Notebook D, 347, Notebook B, 189, 213; Desmond and Moore, *Darwin*, 213–4.

18. Darwin, *The Autobiography*, 98–9.

19. John Macculloch, *Proofs and Illustrations of the Attributes of God* (London: J. Duncan, 1837). It should be noted that Macculloch did put forward an argument that attempted to unify all life-forms in regard to mentality and reason, which also recognized instincts.

20. Darwin, *Charles Darwin's Notebooks*, 633–35, 637.

21. Ibid., Notebook M, 532, 551, Notebook C, 291.

22. Charles Darwin, *The Life and Letters of Charles Darwin*, ed. Francis Darwin (New York: D. Appleton and Company, 1896), 1:384.

23. Desmond and Moore, *Darwin*, 320–23; James Secord, *Victorian Sensation: The Extraordinary Publication, Reception, and Secret Authorship of Vestiges of the Natural History of Creation* (Chicago: University of Chicago Press, 2000).

24. Darwin, *The Autobiography*, 99.

25. Desmond, Moore, and Browne, *Charles Darwin*, 50–62; Gould, *Ever Since Darwin*, 21–27.

26. James Marchant, *Alfred Russel Wallace: Letters and Reminiscences* (New York: Harper & Brothers, 1916), 107–8; Andrew Berry and Janet Browne, "The Other Beetle-Hunter," *Nature* 453:1188–1190.

27. Ross A. Slotten, *The Heretic in Darwin's Court: The Life of Alfred Russel Wallace* (New York: Columbia University Press, 2004), 151–56; Brett Clark and Richard York, "The Restoration of Nature and Biogeography: An Introduction to Alfred Russel Wallace's 'Epping Forest,'" *Organization & Environment* 20/2 (2007): 213–34.

28. Darwin, *The Autobiography*, 101–2.

29. Loren Eiseley, *Darwin's Century* (New York: Anchor Books, 1958); Gould,

Ever Since Darwin; Charles Darwin, *The Origin of Species* (Harmondsworth, England: Penguin Books, 1968).

30. Darwin, *Origin of Species*, 119.

31. Ibid., 459–60; J. W. Burrows, "Introduction," in Darwin, *Origin of Species*, 33.

32. Darwin, *The Autobiography*, 57–8; John Herschel, *A Preliminary Discourse on the Study of Natural Philosophy* (New York: Johnson Reprint Corp., 1966); Darwin, *Origin of Species*, 65.

33. Janet Browne, *Charles Darwin: The Power of Place* (Princeton, N.J.: Princeton University Press, 2002), 107.

34. John Herschel, *Physical Geography of the Globe* (Edinburgh: Adam and Charles Black, 1861), 12; David L. Hull, "Darwin's Science and Victorian Philosophy of Science," in *The Cambridge Companion to Darwin*, ed. Jonathan Hodge and Gregory Radick (Cambridge: Cambridge University Press, 2003), 168–91.

35. Charles Darwin, *The Correspondence of Charles Darwin* (Cambridge: Cambridge University Press, 1994), 9:135.

36. Ibid., 9: 225-27.

37. Darwin, *Life and Letters*, 2:105.

38. Darwin, *Correspondence of Charles Darwin*, 9:200–201.

39. Stephen Jay Gould, "Worm for a Century, and All Seasons," in *Hen's Teeth and Horse's Toes* (New York: W. W. Norton, 1983).

40. Charles Darwin, *The Variation of Animals and Plants Under Domestication* (London: D. Appleton and Company, 1920), 2:415.

41. Charles Darwin, *The Correspondence of Charles Darwin* (Cambridge: Cambridge University Press, 2001), 10:331; Charles Darwin, *The Works of Charles Darwin: The Various Contrivances by which Orchids are Fertilised by Insects* (New York: New York University Press, 1988), vol. 17; Desmond, Moore, and Browne, *Charles Darwin*, 91, 96.

42. Thomas R. Trautmann, *Lewis Henry Morgan and the Invention of Kinship* (Berkeley: University of California Press, 1987), 35, 220.

43. Charles Lyell was a late and reluctant convert to evolutionary theory, and even then he only accepted it in a relatively mild form, starting after 1868. In *The Geological Evidences of the Antiquity of Man* (Philadelphia: G. W. Childs, 1863), he expressly did not endorse evolution, much to Darwin's disappointment. Later, after his conversion, he remained committed to the position that God had preordained the creative steps of evolution, such as that from ape to man.

44. Trautmann, *Lewis Henry Morgan*, 32, 172–73.

45. Charles Darwin, *The Descent of Man, and Selection in Relation to Sex* (Princeton, N.J.: Princeton University Press, 1981), 1:3; John Tyler Bonner and Robert M. May, "Introduction" in ibid., xi. Desmond, Moore, and Browne, *Charles Darwin*, 83–84.

46. Charles Darwin, *The Expression of the Emotions in Man and Animals* (Oxford: Oxford University Press, 1998); Paul Ekman, "Introduction to the Third Edition," in ibid., xxv.

47. Charles Bell, *Anatomy and Philosophy of Expression as Connected with the Fine Arts* (London: Murray, 1844), 3rd ed.; Charles Darwin, *The Expression of the Emotions*; Ekman, "Introduction to the Third Edition," xxxiv.

48. Darwin, *The Autobiography*, 71–72; Desmond and Moore, *Darwin*, 656–57.

49. Darwin, *The Autobiography*, 75–76.

50. Darwin, *The Descent of Man*, 1:65.

51. Ibid., 2:395.

52. Desmond and Moore, *Darwin*, 643–45. Edward B. Aveling was at this time Eleanor Marx's lover, and Darwin's 1880 letter to him, which got mixed up with the Marx family papers, was mistakenly viewed for many years as a letter from Darwin to Karl Marx, from which arose the mistaken legend that Marx had offered to dedicate *Capital* to Darwin. The book that Darwin referred to in his letter was instead Aveling's *The Student's Darwin* (London: Freethought Publishing Company, 1881). See Margaret A. Fay, "Marx and Darwin: A Literary Detective Story," *Monthly Review* 31/10 (1980): 40–57.

53. Quoted in Jeffrey A. Coyne, "Intelligent Design: The Faith That Dare Not Speak Its Name," in *Intelligent Thought: Science Versus the Intelligent Design Movement*, ed. John Brockman (New York: Vintage, 2006), 11.

54. Elizabeth Culotta and Elizabeth Pennisi, "Breakthrough of the Year: Evolution in Action," *Science* 310 (2005): 1878–79; Stephen Jay Gould, *The Structure of Evolutionary Theory* (Cambridge, Mass.: Belknap Press of Harvard University Press, 2002).

55. William A. Dembski, *The Design Revolution* (Downers Grove, Ill.: InterVarsity Press, 2004), 270.

56. Benjamin Wiker, *Moral Darwinism: How We Became Hedonists* (Downers Grove, Ill.: InterVarsity Press, 2002), 220.

57. Ernst Mayr, *The Growth of Biological Thought* (Cambridge, Mass.: Harvard University Press, 1982), 306, 403.

CHAPTER 7: FREUD AND THE ILLUSIONS OF RELIGION

1. Hans Küng, *Freud and the Problem of God* (New Haven: Yale University Press, 1990), 3; Peter Gay, *A Godless Jew: Freud, Atheism, and the Making of Psychoanalysis* (New Haven: Yale University Press, 1987), 53.

2. All three of these thinkers were sharply criticized by Marx and Engels as crude, mechanistic materialists.

3. Küng, *Freud and the Problem of God*, 3–19; Peter Gay, *Freud: A Life in Our Times* (New York: W. W. Norton, 1988), 31. On the biological aspects of Freud's thought, see esp. Frank J. Sulloway, *Freud, Biologist of the Mind* (New York: Basic Books, 1979).

4. Sulloway, *Freud, Biologist of the Mind*, 3–8.

5. Küng, *Freud and the Problem of God*, 3; Erich Fromm, *To Have or To Be* (New York: Harper and Row, 1976), 4; Sulloway, *Freud, Biologist of the Mind*, 4.

6. Peter Gay, "Introduction," in Sigmund Freud, *The Future of an Illusion* (New

York: W. W. Norton, 1989), xi; Küng, *Freud and the Problem of God*, 7–12, 138.

7.   Gay, "Introduction," xxiii; Gay, *Freud*, 602; Sigmund Freud, "A Religious Experience," in *Freud and Freudians on Religion*, ed. Donald Capps (New Haven: Yale University Press, 2001), 59.

8.   Gay, "Introduction," xxiii.

9.   Ernest Jones, *The Life and Work of Sigmund Freud* (New York: Basic Books, 1957), 3:351.

10.   Küng, *Freud and the Problem of God*, 37.

11.   Phillip E. Johnson, "Creator or Blind Watchmaker," in *Intelligent Design Creationism and Its Critics*, ed. R. T. Pennock (Cambridge, Mass.: MIT Press, 2001), 449. Elsewhere Johnson notes that Feuerbach's projection theory of God had a direct influence on both Marx and Freud and uses this to disparage all three. See Johnson, *The Wedge of Truth: Splitting the Foundations of Naturalism* (Downers Grove, Ill.: InterVarsity Press, 2000), 21–22.

12.   Donald De Marco and Benjamin Wiker, *Architects of the Culture of Death* (San Francisco: Ignatius Press, 2004), 15, 209, 218.

13.   Benjamin Wiker and Jonathan Witt, *A Meaningful World* (Downers Grove, Ill.: InterVarsity Press, 2006), 86, 190.

14.   Sigmund Freud, *The Future of an Illusion* (New York: W. W. Norton, 1989), 49.

15.   Sigmund Freud, *Civilization and Its Discontents* (New York: W. W. Norton, 1961), 21; Freud, *Future of an Illusion*, 69.

16.   The other two proofs were "the proof from motion" and "the proof from contingency."

17.   Franz Brentano, *On the Existence of God: Lectures Given at the Universities of Würzburg and Vienna (1868–1891)* (Boston: Martinus Nijhoff Publishers, 1987), 151–267, 290–301, 346; Gay, *Freud*, 29, 31, 526; Peter Gay, *A Godless Jew*, 38, 60; Freud, *Civilization and Its Discontents*, 21. Susan F. Krantz Gabriel, "Brentano on Religion and Natural Theology," in *The Cambridge Companion to Brentano*, ed. Dale Jacquette (Cambridge: Cambridge University Press, 2004), 237–54.

18.   Sigmund Freud, *Moses and Monotheism* (New York: Vintage, 1939), 101, 166.

19.   Sigmund Freud, *Introductory Lectures on Psycho-Analysis*, in the *Standard Edition of the Complete Psychological Works of Sigmund Freud* (London: Hogarth Press, 1963), 15:199.

20.   Sulloway, *Freud, Biologist of the Mind*, 274–75; Stephen Jay Gould, *I Have Landed* (New York: Three Rivers Press, 2003), 147–58; Charles Darwin, *Life and Letters of Charles Darwin* (New York: D. Appleton and Company, 1896), 1:384.

21.   Freud, *Moses and Monotheism*, 127–28; Stephen Jay Gould, *Ontogeny and Phylogeny* (Cambridge, Mass.: Harvard University Press, 1977), 156.

22.   Charles Darwin, *The Descent of Man* (Princeton: Princeton University Press, 1981), 362–63.

23. Küng, *Freud and the Problem of God*, 34–36.
24. Freud, *Totem and Taboo*, 163. The castration of rebellious sons in the primal horde is emphasized in Freud's account in *Moses and Monotheism* rather than *Totem and Taboo*. But there is no doubt that Freud believed it properly belonged to the argument of the former.
25. Sigmund Freud, *Totem and Taboo* (New York: Vintage, 1946), 163, 171, 182–85; Freud, *Moses and Monotheism*, 102–4; Robert A. Paul, *Moses and Civilization: The Meaning Behind Freud's Myth* (New Haven: Yale University Press, 1996), 2–3; Gay, *Freud*, 332.
26. Freud was not of course the first to question the legend of Moses. Voltaire in 1764 raised the question as to whether there was a Moses. See Gay, *Freud*, 606.
27. Freud, *Moses and Monotheism*, 42–43, 119–74; Gay, *Freud*, 606–8.
28. Freud, *Moses and Monotheism*, 144–47.
29. Ibid., 110–14; Paul, *Moses and Civilization*, 3.
30. Freud, *Moses and Monotheism*, 166.
31. Gay, *Freud*, 332–33.
32. Ibid., 647.
33. Gould, *I Have Landed*, 158.
34. See, for example, the argument in Paul, *Moses and Civilization*.
35. Freud, *The Future of an Illusion*, 17–49, 61.
36. Gay, *A Godless Jew*, 35; Sigmund Freud, *New Introductory Lectures on Psychoanalysis* (New York: W. W. Norton, 1965), 150.
37. Freud, *The Future of an Illusion*, 63.
38. Freud, *New Introductory Lectures on Psychoanalysis*, 139–60.
39. Ibid., 152.
40. Ibid., 152–53.
41. Freud, *The Future of an Illusion*, 68–71; Freud, *New Introductory Lectures on Psychoanalysis*, 148.

CHAPTER 8: IN DEFENSE OF NATURAL SCIENCE

1. Karl Marx and Frederick Engels, *Collected Works* (New York: International Publishers, 1975), 1:103. The whole question of the logical relation between design and miracles has presented innumerable problems for design proponents, and special disquisitions on that subject alone. See William Dembski, *Intelligent Design: The Bridge Between Science and Theology* (Downers Grove, Ill.: InterVarsity Press, 1999), 49–69.
2. The intelligent design movement's attack on science has not been limited only to the criticism of evolutionary theory, but also chemistry, physics, cosmology, and other scientific fields. Here we focus our discussion simply on intelligent design's challenge to evolutionary biology. The intelligent design criticisms of other sciences are if anything even less developed than their attack on evolutionary biology. In all these areas, as in their response to Darwinism, as Robert Pennock writes, "anti-evolutionists avoid the hopeless task of giving positive

evidence for their preferred view [i.e., an Intelligent Designer] by trying to poke holes in the scientific account." The result is a set of arguments that in line with the traditional "God of the Gaps" argument point to gaps in current scientific understandings of the origins of life, thermodynamics, cosmology, etc., followed by the assertion that only an Intelligent Designer can fill these gaps. For example, the intelligent design textbook for junior high and high school students, *Of Pandas and People*, which Discovery Institute fellows helped prepare, has a chapter on the origins of life, which, after pointing to some of the issues that scientists have been struggling with through numerous experimental investigations, ends with the essentially meaningless statement: "What makes this interpretation [intelligent design] so compelling is the amazing correlation between the structure of information molecules (DNA, protein) and our universal experience that such sequences are the result of intelligent causes." From a scientific standpoint, such claims are of course completely empty of content, and, what is more, science stoppers. See Robert T. Pennock, "God of the Gap," in *Scientists Confront Intelligent Design and Creationism*, ed. Andrew J. Petto and Laurie R. Godfrey (New York: W. W. Norton, 2007), 324; Niall Shanks, *God, the Devil, and Darwin* (Oxford: Oxford University Press, 2006); Percival Davis and Dale H. Kenyon (Charles B. Thaxton, academic editor), *Of Pandas and People* (Dallas: Haughton Publishing, 1993), 58.

3.   Richard Lewontin and Richard Levins, *Biology Under the Influence* (New York: Monthly Review Press, 2007), 92.

4.   It is true that some intelligent design proponents, most notably Behe, formally argue that the "intelligent designer" could be some other natural entity, perhaps "made of gas particles," and possibly could "begin as clay crystals." There are serious scientists, we are told, who believe that "life could exist in such places as the atmosphere of Jupiter." But it is hard to take such notions of possible naturalistic intelligent designers seriously, since the "Wedge" strategy of intelligent design proponents has always been structured around promoting the notion of God (particularly the God of fundamentalist Christianity) and is rooted in an attack on "naturalism." An extraterrestrial natural being of some sort would therefore hardly meet the needs of this intellectual tradition, which is aimed at positing the supernatural. Behe, meanwhile, makes no bones about his own intellectual motivation as a "theist." Michael Behe, "Reply to My Critics," *Biology and Philosophy* 16 (November 2001), 698–99, 702, 705; Dembski, *Intelligent Design*, 78–79, 187–210.

5.   Stephen C. Meyer, "Not By Chance," *National Post*, December 1, 2005.

6.   David Hume, *Enquiries Concerning Human Understanding and Concerning the Principles of Morals* (Oxford: Oxford University Press, 1975), 142. Hume's position here was discussed more fully in chapter 4 above.

7.   Behe, "Reply to My Critics," 703.

8.   Michael Behe, *Darwin's Black Box* (New York: Free Press, 1996), 39. See also the discussion in chapter 1 of the present volume.

9.   Davis and Kenyon, *Of Pandas and People*, 15–16.

10.   John G. West, *Darwin Day in America* (Wilmington, Del.: Intercollegiate Studies Institute Books, 2007), 232. See also Phillip E. Johnson, *The Wedge of Truth: Splitting the Foundations of Naturalism* (Downers Grove, Ill.: InterVarsity Press, 2000), 127–31.

11.   See Dembski, *Intelligent Design*, 86.

12.   Kenneth P. Miller, *Only a Theory* (New York: Viking, 2008), 58–62.

13.   Behe, "Reply to My Critics," 695.

14.   Michael Behe, *The Edge of Evolution: The Search for the Limits of Darwinism* (New York: Free Press, 2007).

15.   Mark J. Pallen and Nicholas J. Matzke, "From *The Origin of Species* to the Origin of Bacterial Flagella," *Nature Reviews Microbiology* 4/10 (2006): 784–90; also Paul Davies, *The Goldilocks Enigma* (Boston: Houghton Mifflin, 2006), 193–96. For an insightful and useful critique of Behe and intelligent design in general, see Sahotra Sarkar, *Doubting Darwin?* (Oxford: Blackwell, 2007); U.S. District Court for the Middle District of Pennsylvania, *Kitzmiller v. Dover Area School District et al.*, http://www.pamd.uscourts.gov/kitzmiller/kitzmiller_342.pdf, 74–78.

16.   Behe, *The Edge of Evolution*.

17.   William A. Dembski, *The Design Inference* (Cambridge: Cambridge University Press, 1998).

18.   Dembski, *Intelligent Design*, 165.

19.   See Brett Clark and Richard York, "Dialectical Materialism and Nature," *Organization & Environment* 18/3 (2005): 318–37; John Bellamy Foster, *Marx's Ecology* (New York: Monthly Review Press, 2000).

20.   Richard Dawkins, *The Blind Watchmaker: Why the Evidence of Evolution Reveals a Universe Without Design* (New York: W. W. Norton, 1986), 81.

21.   Stephen Jay Gould, *The Structure of Evolutionary Theory* (Cambridge, Mass.: Belknap Press of Harvard University Press, 2002); Stephen Jay Gould, "Is a New and General Theory of Evolution Emerging?" *Paleobiology* 6/1 (1980): 119–30.

22.   Gould, "Is a New and General Theory of Evolution Emerging?" 129; Richard Lewontin, *The Triple Helix: Gene, Organism, and Environment* (Cambridge, Mass.: Harvard University Press, 2000).

23.   Stephen Jay Gould, *Eight Little Piggies* (New York: W. W. Norton, 1993); Gould, "Is a New and General Theory of Evolution Emerging?" 129.

24.   Gould, *The Structure of Evolutionary Theory*.

25.   See, for example, the explicit support of structuralism and the views of Owen and Saint-Hilaire by Benjamin Wiker and Jonathan Witt, *A Meaningful World* (Downers Grove, Ill.: InterVarsity Press, 2006), 229.

26.   Gould, *The Structure of Evolutionary Theory*, 251–60.

27.   Stephen Jay Gould and Richard Lewontin, "The Spandrels of San Marco and the Panglossian Paradigm: A Critique of the Adaptationist Program,"

Proceedings of the Royal Society of London. Series B, Biological Sciences 205/1161 (1979): 581–98.

28.   Gould, *The Structure of Evolutionary Theory*, 1259–60.

29.   Stephen Jay Gould and Elisabeth S. Vrba, "Exaptation—A Missing Term in the Science of Form," *Paleobiology* 8 (1982): 4–15.

30.   Stephen Jay Gould, *Bully for Brontosaurus* (New York: W. W. Norton, 1991), 139–51.

31.   Gould, *The Structure of Evolutionary Theory*; Steven Rose, *The Future of the Brain: The Promise and Perils of Tomorrow's Neuroscience* (Oxford: Oxford University Press, 2005); Ian Tattersall, *Becoming Human: Evolution and Human Uniqueness* (New York: Harcourt Brace, 1998).

32.   Pallen and Matzke, "From *The Origin of Species*."

33.   Behe, *Darwin's Black Box*, 39.

34.   D'Arcy Wentworth Thompson, *On Growth and Form: A New Edition* (Cambridge: Cambridge University Press, 1942).

35.   Stuart A. Kauffman, *The Origins of Order: Self-Organization and Selection in Evolution* (New York: Oxford University Press, 1993).

36.   *The Origin of Order*, 482.

37.   Thomas S. Hall, *Ideas of Life and Matter: Studies in the History of General Physiology, 1600 B.C.–1900 A.D.* (Chicago: University of Chicago Press, 1969), 1:19–20.

CHAPTER 9: REPLAYING THE TAPE OF LIFE

1.   See Stephen Jay Gould, *Ever Since Darwin* (New York: W. W. Norton, 1977); Brett Clark and Richard York, "Dialectical Materialism and Nature," *Organization & Environment* 18/3 (2005): 318–37.

2.   Niles Eldredge and Stephen Jay Gould, "Punctuated Equilibria: An Alternative to Phyletic Gradualism," in *Models of Paleobiology*, ed. T. J. M. Schopf (San Francisco: Freeman, Cooper and Co., 1972), 82–115; Stephen Jay Gould and Niles Eldredge, "Punctuated Equilibria: The Tempo and Mode of Evolution Reconsidered," *Paleobiology* 3 (1977): 115–51.

3.   Phillip E. Johnson, *Darwin on Trial* (Washington, D.C.: Regnery Gateway, 1991), 61–62.

4.   Stephen Jay Gould, *Wonderful Life* (New York: W. W. Norton, 1989).

5.   Stephen Jay Gould, "The Persistently Flat Earth: Irrationality and Dogmatism Are Foes of Both Science and Religion," *Natural History* 103/3 (1994): 12–9; David M. Raup, *Extinction: Bad Genes or Bad Luck?* (New York: W. W. Norton, 1991).

6.   For an assessment of the potential causes of the Permian extinction, see Douglas H. Erwin's *Extinction: How Life on Earth Nearly Ended 250 Million Years Ago* (Princeton, N.J.: Princeton University Press, 2006).

7.   Stephen Jay Gould, *Full House* (New York: Harmony Books, 1996).

8.   There were, of course, organic chemical precursors to the single-celled organ-

ism, as there are organic replicators, such as viruses, that evolved subsequently, but these are typically not considered sufficiently structured to be called "life."

9. Gould, *Full House*.

10. Ibid., 175–95.

11. See Richard Lewontin and Richard Levins, *Biology Under the Influence* (New York: Monthly Review Press, 2007), 92.

12. Stephen Jay Gould, *The Panda's Thumb* (New York: W. W. Norton, 1980), 20–21.

13. William Dembski, *Intelligent Design* (Downers Grove, Ill.: InterVarsity Press, 1999), 261–64, 286, 301; and "What Intelligent Design Is Not," in *Signs of Intelligence*, ed. William A. Dembski and James M. Kushiner (Grand Rapids, Mich.: Brazos Press, 2001), 7–12.

14. Richard Levins and Richard Lewontin, *The Dialectical Biologist* (Cambridge, Mass.: Harvard University Press, 1985), 286–88.

15. Levins and Lewontin, *Dialectical Biologist*, 136; Richard Lewontin, *The Triple Helix: Gene, Organism, and Environment* (Cambridge, Mass.: Harvard University Press, 2000).

16. Levins and Lewontin, *Dialectical Biologist*, 277, 288.

17. Gould, *Wonderful Life*.

CHAPTER 10: THE END OF THE WEDGE

1. Darwin quoted in Stephen Jay Gould, *Eight Little Piggies* (New York: W. W. Norton, 1993), 302; Charles Darwin, *On the Origin of Species* (Cambridge, Mass.: Harvard University Press, 1964), 67; Charles Darwin, *Charles Darwin's Notebooks, 1836–1844* (Ithaca, N.Y.: Cornell University Press, 1987), 375–76.

2. Gould, *Eight Little Piggies*, 303–12. Gould argues that Darwin used the wedge metaphor uncharacteristically to point to a kind of direction and even progress to natural selection in conformity with the prejudices of the Victorian era in which he lived. More generally, Darwin rejected any suggestion of a direction (or purpose) to natural processes. See Stephen Jay Gould, *Ever Since Darwin* (New York: W. W. Norton, 1977), 12–13.

3. Phillip E. Johnson, *Defeating Darwinism* (Downers Grove, Ill.: InterVarsity Press, 1997), 92; Barbara Forrest and Paul R. Gross, *Creationism's Trojan Horse: The Wedge of Intelligent Design* (Oxford: Oxford University Press, 2007), 22.

4. William A. Dembski, "Dealing with the Backlash Against Intelligent Design," in *Darwin's Nemesis*, ed. William A. Dembski (Downers Grove, Ill.: InterVarsity Press, 2006), 100.

5. William A. Dembski, *The Design Revolution* (Downers Grove, Ill.: InterVarsity Press, 2004), 306–9; Niall Shanks, *God, the Devil, and Darwin* (Oxford: Oxford University Press, 2006), 11–12, 157–59, 258; Eugenie C. Scott, *Evolution vs. Creationism* (Berkeley: University of California Press, 2004), 125–26.

6.   Phillip E. Johnson, *The Wedge of Truth: Splitting the Foundations of Naturalism* (Downers Grove, Ill.: InterVarsity Press, 2000), 14.

7.   Center for Renewal of Science and Culture, Discovery Institute, *The Wedge Strategy*, http://www.antievolution.org/features/wedge.html; Forrest and Gross, *Creationism's Trojan Horse*, 30; John R. Cole, "Wielding the Wedge," in *Scientists Confront Intelligent Design and Creationism*, ed. Andrew J. Petto and Laurie R. Godfrey (New York: W. W. Norton, 2007), 110–28.

8.   Center for the Renewal of Science and Culture, quoted in Scott, *Evolution vs. Creationism*, 124–25.

9.   William A. Dembski, *Intelligent Design: The Bridge Between Science and Theology* (Downers Grove, Ill.: InterVarsity Press, 1999), 107, 246–52.

10.   William A. Dembski, "Signs of Intelligence," in *Signs of Intelligence*, ed. William A. Dembski and James M. Kushiner (Grand Rapids, Michigan: Brazos Press, 2001), 192.

11.   Dembski, *Intelligent Design*, 102–3, 187–210, 225–30.

12.   Ibid., 191–93; Johnson quoted in Edward Humes, *Monkey Girl* (New York: HarperCollins, 2007), 71.

13.   Center for Renewal of Science and Culture, Discovery Institute, *The Wedge Strategy*, http://www.antievolution.org/features/wedge.html; Jay Wesley Richards, "Proud Obstacles and a Reasonable Hope," in Dembski and Kushiner, *Signs of Intelligence*, 51–59.

14.   Dembski, *Intelligent Design*, 261–64; Dembski, "Introduction," in Dembski and Kushiner, eds., *Signs of Intelligence*, 7–12; Jonathan Witt and Benjamin Wiker, *The Meaningful World* (Downers Grove, Ill.: InterVarsity Press, 2006), 28–29.

15.   John G. West, *Darwin Day in America* (Wilimington, Del.: Intercollegiate Studies Institute, 2007), 6–8. In his relatively sophisticated treatment of the Epicurean tradition, West acknowledges that Epicurus's materialism did not contradict free will.

16.   John G. West, *Darwin Day in America*, 41–42, 118–22, 216.

17.   West. *Darwin Day in America*, ix–x, 375; John G. West, "The Regeneration of Science and Culture: The Cultural Implications of Scientific Materialism Versus Intelligent Design," in Dembski and Kushiner, *Signs of Intelligence*, 60–69; Shanks, *God, the Devil, and Darwin*, 226–29.

18.   John G. West, "C. S. Lewis and Materialism," *Religion and Liberty* 6/6 (November–December 1996): 6–8.

19.   Benjamin Wiker, *Moral Darwinism: How We Became Hedonists* (Downers Grove, Ill.: InterVarsity Press, 2002); and "Darwin as Epicurean: Interview," *Touchstone* 15/8 (October 2002), http://www.touchstone.mag/; and *Ten Books That Screwed the World: And Five Others That Didn't Help* (Washington, D.C.: Regnery Publishing, 2008), 147; Donald De Marco and Benjamin Wiker, *Architects of the Culture of Death* (San Francisco: Ignatius Press, 2004). The claim that Darwin helped pave the way to Hitler is not advanced by Wiker alone

but is also promoted by Center for Science and Culture fellow Richard Weikart in his *From Darwin to Hitler* (New York: Palgrave Macmillan, 2004). Fuller, basing himself on Weikart's book, points to Darwinism, environmental sociology, and human ecology as somehow related to "racial hygiene" and "blood and soil" approaches to science and culture. See Steve Fuller, *Science vs. Religion: Intelligent Design and the Problem of Evolution* (Cambridge: Polity Press, 2007), 149.

20. Witt and Wiker, *The Meaningful World*, 16, 60.

21. Ibid., 108–9, 247–49; Wiker, *Moral Darwinism*, 296–97.

22. Wiker and Witt, *The Meaningful World*, 248.

23. Karl Marx and Frederick Engels, *Collected Works* (New York: International Publishers, 1975), 5:141–42, emphasis added.

24. James, 3:13–14, 4:4. See "The General Epistle of St. James" (trans. John Wesley), Wesley Center Online, http://wesley.nnu.edu/john_wesley/Wesley_NT/20-Jam.html; J. B. S. Haldane, "Auld Hornie, F.R.S.," *The Modern Quarterly*, no. 4 (Autumn 1946): 39–40; Helen Ellerbe, *The Dark Side of Christianity* (Windermer, Florida: Morningstar and Lark, 1995), 140.

25. C. S. Lewis, *Miracles* (New York: Harper and Row, 1960), 6, 27–28.

26. C. S. Lewis, *The Abolition of Man* (New York: HarperCollins, 1974); *Out of the Silent Planet* (New York: Macmillan, 1971); *The Seeing Eye* (New York: Ballantine, 1967), 113–29; and *God in the Dock* (Grand Rapids, Mich.: Eerdmans, 1970), 208–11; Jeffrey D. Schultz and John G. West, eds., *The C. S. Lewis Readers' Encyclopedia* (Grand Rapids, Mich,: Zondervan Publishing House, 1998), 158–59, 178–79, 308–9, 316–17, 360–61, 396–99, 425; J. B. S. Haldane, "More Anti-Lewisite," http://www.solcon.nl/arendsmilde/cslewis/reflections/e-haldane.htm#Anti-Lewisite.

27. Lewis, *The Seeing Eye*, 121.

28. Charles S. Peirce, *Essays in the Philosophy of Science* (New York: Bobbs-Merrill, 1957). For William James's critique of the argument from design, see James, *Varieties of Religious Experience* (New York: Modern Library, 2002), 535–38. Nancy Pearcey attacks Peirce from an intelligent design standpoint, both for the influence of Epicurus on his thought and his role as a "Darwinian of the mind." See Nancy Pearcey, *Total Truth: Liberating Christianity From Its Cultural Captivity* (Wheaton, Ill.: Good News Publications, 2004), 235–36, 389. On Peirce and Epicurus, see David R. Suits, "The Fixation of Satisfaction: Epicurus and Peirce on the Goal," in *Epicurus: His Continuing Influence and Contemporary Relevance*, ed. Diana R. Gordon and David R. Suits (Rochester, N.Y.: Rochester Institute of Technology, Cary Graphics Arts Press, 2003), 139–55.

29. Johnson, *The Wedge of Truth*, 86–87, 91–92, 99–101.

30. Wiker, *Moral Darwinism*, 314–15.

31. Fuller, *Science vs. Religion*, 1–3.

32. Stephen Jay Gould, *Ever Since Darwin* (New York: W. W. Norton, 1977), 12–13.

33. Karl Marx, *Early Writings* (London: Penguin, 1974), 328. Ironically, Dembski, in a retort aimed at Marx, claims that belief in Christ precludes human alienation, since Christ is both man and God. "Christ himself," he writes, "was human. Karl Marx's favorite maxim was *Nihil humani a me alienum puto* (Nothing human is alien to me). The Christian has more right than anyone to this maxim." Alienation, however, is viewed by Dembski here in Christian spiritualist terms, and not as alienation from human sensuous being (that is, in terms of human beings as natural and social beings), as in Marx. See Dembski, *Intelligent Design*, 206-07.

34. Richard Lewontin, "Billions and Billions of Demons," *New York Review of Books* 64/1 (1997): 28–32.

35. See István Mészáros, *Marx's Theory of Alienation* (London: Merlin Press, 1971), 162–89.

36. Benedict Spinoza, *Ethics* (London: Penguin, 1996), 29; Marx and Engels, *Collected Works*, 1:30.

INDEX